CAMBRIDGE TRACTS IN MATHEMATICS

General Editors

B. BOLLOBAS, F. KIRWAN, P. SARNAK, C.T.C. WALL

135 Solitons

T. Miwa
Research Institute for Mathematical Sciences
Kyoto University

M. Jimbo
Kyoto University

E. Date
University of Osaka

Translated by Miles Reid

Solitons:
Differential Equations, Symmetries and Infinite Dimensional Algebras

PUBLISHED BY THE PRESS SYNDICATE OF THE UNIVERSITY OF CAMBRIDGE
The Pitt Building, Trumpington Street, Cambridge, United Kingdom

CAMBRIDGE UNIVERSITY PRESS
The Edinburgh Building, Cambridge CB2 2RU, UK   http://www.cup.cam.ac.uk
40 West 20th Street, New York, NY 10011–4211, USA   http://www.cup.org
10 Stamford Road, Oakleigh, Melbourne 3166, Australia
Ruiz de Alarcón 13, 28014 Madrid, Spain

Soliton no Suri (Mathematics of Solitons)
by Tetsuji Miwa, Michio Jimbo and Etsuro Date
Copyright © Tetsuji Miwa, Michio Jimbo and Etsuro Date
Originally published in Japanese by Iwanami Shoten, Publishers, Tokyo in 1993.
English translation © Cambridge University Press 2000

This book is in copyright. Subject to statutory exception
and to the provisions of relevant collective licensing agreements,
no reproduction of any part may take place without
the written permission of Cambridge University Press.

First published in English 2000

Printed in the United Kingdom at the University Press, Cambridge

Typeface Computer Modern 10/11pt   *System* LaTeX2e   [UPH]

A catalogue record for this book is available from the British Library

ISBN 0 521 56161 2 hardback

Contents

Preface		*page* vii
1	The KdV equation and its symmetries	1
1.1	Symmetries and transformation groups	1
1.2	Symmetries of the KdV equation	5
1.3	The Lax form of an evolution equation (the approach via linear differential equations)	8
Exercises to Chapter 1		9
2	The KdV hierarchy	11
2.1	Pseudodifferential operators	11
2.2	Higher order KdV equations	13
2.3	Infinitely many commuting symmetries	14
2.4	The KP hierarchy	16
Exercises to Chapter 2		18
3	The Hirota equation and vertex operators	19
3.1	The Hirota derivative	19
3.2	n-Solitons	22
3.3	Vertex operators	24
3.4	The bilinear identity	28
Exercises to Chapter 3		31
4	The calculus of Fermions	32
4.1	The Bosonic algebra of differentiation and multiplication	32
4.2	Fermions	34
4.3	The Fock representation	35
4.4	Duality, charge and energy	37
4.5	Wick's theorem	40
Exercises to Chapter 4		41
5	The Boson–Fermion correspondence	43
5.1	Using generating functions	43

5.2	The normal product	44
5.3	Realising the Bosons	46
5.4	Isomorphism of Fock spaces	47
5.5	Realising the Fermions	49
Exercises to Chapter 5		52
6	Transformation groups and tau functions	53
6.1	Group actions and orbits	53
6.2	The Lie algebra $\mathfrak{gl}(\infty)$ of quadratic expressions	54
6.3	The transformation group of the KP hierarchy	58
Exercises to Chapter 6		60
7	The transformation group of the KdV equation	61
7.1	KP hierarchy versus KdV hierarchy	61
7.2	The Transformation group of the KdV equation	63
Exercises to Chapter 7		65
8	Finite dimensional Grassmannians and Plücker relations	66
8.1	Finite dimensional Grassmannians	66
8.2	Plücker coordinates	69
8.3	Plücker relations	71
Exercises to Chapter 8		75
9	Infinite dimensional Grassmannians	76
9.1	The case of finite dimensional Fock space	76
9.2	Description of the vacuum orbit	80
9.3	Young diagrams and character polynomials	82
Exercises to Chapter 9		87
10	The bilinear identity revisited	88
10.1	The bilinear identity and the Plücker relations	88
10.2	Plücker relations and the Hirota equation	90
Exercises to Chapter 10		93
Solutions to exercises		94
Bibliography		103
Index		107

Preface

Waves and vibrations are among the most basic forms of motion, and their study goes back a very long way. Small amplitude waves are described mathematically by a linear differential equation, and their behaviour can be studied in detail. In contrast, when the amplitude is not restricted to being small, the differential equation becomes nonlinear, and its analysis becomes in general an extremely difficult problem.

An example of a nonlinear wave equation is the model

$$\frac{\partial u}{\partial t} + 6u\frac{\partial u}{\partial x} + \frac{\partial^3 u}{\partial x^3} = 0 \tag{1}$$

for a shallow water wave. This equation was proposed by the physicists Korteweg and de Vries at the end of the nineteenth century, and is now called the KdV equation. In particular, if we now assume a solution in the form of a travelling wave $u(x,t) = f(x - ct)$ then (1) can be integrated: imposing the boundary conditions at large distances that $u(x,t)$ tends to 0 sufficiently fast as $x \to \pm\infty$, we find the exact solution

$$u_1(x,t) = \frac{c}{2}\text{sech}^2\left(\frac{\sqrt{c}}{2}(x - ct + \delta)\right), \tag{2}$$

where δ is a constant of integration. The motion described by this is an isolated wave, localised in a small part of space. In fact, in addition to this solution, (1) is known to have an infinite series of exact solutions $u_2(x,t), u_3(x,t), \ldots$.

These solutions $u_n(x,t)$ contain $2n$ arbitrary parameters c_i, δ_i, and in the distant past $t \ll 0$ and the distant future $t \gg 0$ they behave just like a superposition of independent isolated waves of the form (2). The isolated waves can overtake or collide with one another in finite time, but they revert after the collision to their individual independent state (except for a possible phase change), and are transmitted without annihilation. A wave motion with this special type of particle-like behaviour is called a *soliton*, and a solution $u_n(x,t)$ representing n isolated waves is called an *n-soliton*.

For a linear differential equation, the *principle of superposition* says that if special solutions u_i are known for $i = 1, \ldots, N$, then other solutions containing arbitrary constants can be made as linear combinations $\sum_{i=1}^{N} c_i u_i$. The principle of superposition does not apply to the KdV equation, because it is nonlinear. The fact that, despite this, there exist exact solutions containing an arbitrary number of parameters, is a remarkable and exceptional phenomenon, and it suggests that the KdV equation occupies a special position among general nonlinear differential equations.

In classical mechanics, there is a notion of a *completely integrable system* (we say simply *integrable system* for short): consider a mechanical system with f degrees of freedom

$$\frac{dq_i}{dt} = \frac{\partial H}{\partial p_i}, \quad \frac{dp_i}{dt} = -\frac{\partial H}{\partial q_i} \quad \text{for } i = 1, \ldots, f. \tag{3}$$

Here H is a Hamiltonian. We say that (3) is a completely integrable system if it has f independent first integrals $F_1(q,p) = H(p,q), \ldots, F_f(q,p)$. When this holds, the general solution of (3) can be obtained by solving $F_i(q,p) = C_i$ for $i = 1, \ldots, f$, where C_i are arbitrary constants. Now it is known the KdV equation can in fact be interpreted as an integrable system in this sense, but having infinitely many degrees of freedom. The existence of infinitely many exact solutions such as the soliton solutions is a reflection of this complete integrability.

Although these remarkable properties of the KdV equation were considered as an isolated special phenomenon when they were first discovered, their universal nature became gradually more apparent in rapidly developing research from the late 1960s onwards. At present, a huge number of concrete examples of integrable nonlinear differential (and difference) equations are known. These are also quite generally called soliton equations. A model example of these is the Toda lattice discovered by Morikazu TODA. Many techniques for finding exact solutions of these equations have also been discovered: inverse scattering theory which solves the initial value problem, the bilinear method initiated by Ryogo HIROTA, the theory of quasiperiodic solutions based on Riemann surfaces and theta functions, etc.

At the same time, classical results that had remained long buried came to be viewed in a new light: the applications of theta functions to classical mechanics, nonlinear differential equations arising in the differential geometry of surfaces, the study of commutative subrings of rings of differential operators, and so on. One could say that the KP (Kadomstev–Petviashvili) equations, which generalise the KdV equations, the Toda equation, the Hirota derivative and so on, had already existed in a different form. The theory of integrable systems was confirmed as a paradigm providing a unified viewpoint on these various results.

What is the guiding principle behind the complete integrability of all these systems? In a word, it is the extremely high degree of *symmetry* hidden in the system. For 'high degree of symmetry', we could equally well say 'action of a huge transformation group'. The aim of this book is to use the KdV and KP equations as material to introduce the idea of an infinite dimensional transformation group acting on spaces of solutions of integrable systems. Mikio SATO discovered that the totality of solutions of the KP equations form an infinite dimensional Grassmannian, and established the algebraic structure theory of completely integrable systems. Our aim is to explain the essence of this theory of Sato, together with development of these ideas in the research of Masaki KASHIWARA and the present authors, without going into all the details. We leave to the reader's kind judgment the extent to which we have succeeded in our aim.

As far as prerequisites are concerned, we have tried to write the book so that it can be read by a student with a knowledge of differential and integral calculus, linear algebra and elementary complex analysis (up to the calculus of residues).

Finally, we would like to thank Shigeki SUGIMOTO, Takeshi SUZUKI and Masato HAYASHI for reading the manuscript and making useful comments.

Tetsuji MIWA
Michio JIMBO
Etsuro DATE

Kyoto and Osaka, 1992

1
The KdV equation and its symmetries

We look for symmetries of the KdV equation taking the form of infinitesimal transformations by a nonlinear evolution equation. The KdV equation is itself a nonlinear evolution equation, but we will see how to derive it in terms of compatibility conditions between *linear* equations.

The best possible compass to guide us in mathematics and the natural sciences is the notion of *symmetry*. Following this compass, up anchor and away over the wide ocean of solitons!

1.1 Symmetries and transformation groups

So then, what is symmetry? For example, consider the symmetries of the circle. One sees fairly intuitively that the circle is taken into itself by either

(1) a rotation around the centre, or
(2) a reflection in a diameter.

How do we express this intuition in precise mathematical terms? In the (x, y) coordinate plane, the circle is given as the set of points satisfying

$$x^2 + y^2 = r^2. \tag{1.1}$$

A rotation of the plane is the linear transformation

$$\begin{pmatrix} x' \\ y' \end{pmatrix} = \begin{pmatrix} \cos\theta & -\sin\theta \\ \sin\theta & \cos\theta \end{pmatrix} \begin{pmatrix} x \\ y \end{pmatrix}, \tag{1.2}$$

and a reflection

$$\begin{pmatrix} x' \\ y' \end{pmatrix} = \begin{pmatrix} \cos\theta & \sin\theta \\ \sin\theta & -\cos\theta \end{pmatrix} \begin{pmatrix} x \\ y \end{pmatrix}. \tag{1.3}$$

The linear transformation given by

$$\begin{pmatrix} x' \\ y' \end{pmatrix} = \begin{pmatrix} a & b \\ c & d \end{pmatrix} \begin{pmatrix} x \\ y \end{pmatrix}, \qquad (1.4)$$

represents a symmetry of the circle if it preserves the expression (1.1). In other words, we say that (1.4) is a *symmetry* of (1.1) if (x', y') is a solution of (1.1) whenever (x, y) is.

Write $T(\theta)$ for the transformation (1.2), and $S(\theta)$ for (1.3). The set of all invertible linear transformations forms a group under composition. In other words, if we define the product of two matrices $T_1 = \begin{pmatrix} a_1 & b_1 \\ c_1 & d_1 \end{pmatrix}$ and $T_2 = \begin{pmatrix} a_2 & b_2 \\ c_2 & d_2 \end{pmatrix}$ with nonzero determinant to be the matrix $T_1 \cdot T_2 = \begin{pmatrix} a_1 & b_1 \\ c_1 & d_1 \end{pmatrix} \begin{pmatrix} a_2 & b_2 \\ c_2 & d_2 \end{pmatrix}$ then the group axioms† are satisfied:

(1) Associativity: $(T_1 \cdot T_2) \cdot T_3 = T_1 \cdot (T_2 \cdot T_3)$.
(2) Existence of the unit: $T \cdot \text{id} = \text{id} \cdot T$; here $\text{id} = \begin{pmatrix} 1 & 0 \\ 0 & 1 \end{pmatrix}$.
(3) Existence of the inverse: $T \cdot T^{-1} = T^{-1} \circ T = \text{id}$.

Restricting attention in particular to the elements that leave the circle (1.1) invariant, these also form a group. This is called a *transformation group* of the circle. We have $T(\theta) = T(\theta')$, or $S(\theta) = S(\theta')$, if and only if $\theta = \theta' + 2n\pi$, where n is an integer. The group law is given by

$$\left. \begin{array}{l} T(\theta_1) \cdot T(\theta_2) = T(\theta_1 + \theta_2), \\ T(\theta_1) \cdot S(\theta_2) = S(\theta_2) \cdot T(-\theta_1) = S(\theta_1 + \theta_2), \\ S(\theta_1) \cdot S(\theta_2) = T(\theta_1 - \theta_2), \end{array} \right\} \qquad (1.5)$$

By passing from the transformations themselves to the composition rules (1.5), the symmetries of the circle are isolated from their concrete nature as transformations of the plane, and purified into an abstract group law.

Among symmetries of the circle, consider only the rotations $T(\theta)$. When the parameter θ is 0 we have $T(0) = \text{id}$, so that we can view the transformation

$$\begin{pmatrix} x(\theta) \\ y(\theta) \end{pmatrix} = \begin{pmatrix} \cos\theta & -\sin\theta \\ \sin\theta & \cos\theta \end{pmatrix} \begin{pmatrix} x \\ y \end{pmatrix}, \qquad (1.6)$$

varying together with θ, as the process which takes a given solution (x, y) of the algebraic equation (1.1) continuously around the circle.

† See any textbook on group theory, for example W. Ledermann, *Introduction to group theory*, Oliver and Boyd, 1973, or P.M. Cohn, *Algebra*, Vol. I, Wiley, 1974, Section 3.2.

1.1 Symmetries and transformation groups

Differentiating this with respect to θ gives

$$\frac{d}{d\theta}\begin{pmatrix}x(\theta)\\y(\theta)\end{pmatrix}=\begin{pmatrix}0&-1\\1&0\end{pmatrix}\begin{pmatrix}x(\theta)\\y(\theta)\end{pmatrix}. \tag{1.7}$$

The transformation (1.6) is completely determined by these equations, together with the initial condition

$$\begin{pmatrix}x(0)\\y(0)\end{pmatrix}=\begin{pmatrix}x\\y\end{pmatrix}. \tag{1.8}$$

The operator $\begin{pmatrix}0&-1\\1&0\end{pmatrix}$ is an *infinitesimal generator* of the rotation, in a sense we explain presently. We have the relation

$$T(\theta)=e^{\theta\begin{pmatrix}0&-1\\1&0\end{pmatrix}}.$$

Expanding this with θ as a small parameter gives

$$T(\theta)=1+\theta\begin{pmatrix}0&-1\\1&0\end{pmatrix}+O(\theta^2). \tag{1.9}$$

More generally, if $R(\theta)$ is a transformation depending on one parameter θ and satisfying $R(\theta_1+\theta_2)=R(\theta_1)R(\theta_2)$, and we have $R(\theta)=1+\theta X+O(\theta^2)$, then we say that X is the *infinitesimal generator* of the one parameter transformation $R(\theta)$, and we set $R(\theta)=e^{\theta X}$. (Compare the discussion of Lie algebras at the end of this section.) If we think of $R(\theta)$ as acting on an initial object f, and we write $f^R(\theta)=R(\theta)f$, then

$$\frac{df^R}{d\theta}=\frac{d}{d\theta}R(\theta)f=Xf^R(\theta). \tag{1.10}$$

We are primarily interested in transformations and infinitesimal transformations acting on functions. For example, for a function of two variables $f(x,y)$, consider the differential equation

$$\left(\frac{\partial^2}{\partial x^2}+\frac{\partial^2}{\partial y^2}-r^2\right)f(x,y)=0, \tag{1.11}$$

where r is a constant, independent of (x,y). Whereas for the algebraic equation (1.1) we looked for solutions in the 2 dimensional (x,y) plane, for the differential equation (1.11), the solution $f(x,y)$ should live in the infinite dimensional vector space of functions of two variables. Now a rotation of the (x,y) plane induces an action $g\mapsto T(\theta)g$ on the space of functions g of two variables (x,y) by the formula

$$(T(\theta)g)(x,y)=g(x(-\theta),y(-\theta))$$

(compare (1.6)). Or by considering $f^T(x,y;\theta) = (T(\theta)f)(x,y)$, we can write this as an infinitesimal transformation, giving

$$\frac{\partial}{\partial \theta} f^T(x,y;\theta) = \left(x\frac{\partial}{\partial y} - y\frac{\partial}{\partial x}\right) f^T(x,y;\theta).$$

Thus the operator $x(\partial/\partial y) - y(\partial/\partial x)$ is an infinitesimal generator of the transformation. The equation (1.11) has rotational symmetry; thus if f is a solution of (1.11), so is $T(\theta)f$. The equation (1.11) is also invariant under parallel translation $(x,y) \mapsto (x+a, y+b)$. Expressing parallel translation also as an infinitesimal transformation gives

$$f(x+a, y+b) = e^{a\frac{\partial}{\partial x} + b\frac{\partial}{\partial y}} f(x,y), \qquad (1.12)$$

which is just the Taylor expansion of f around (x,y). (We note here that this equation is often used in what follows. Compare also Exercise 1.1.)

For use in future chapters, we now want to give a brief treatment of Lie algebras. Suppose that X and Y are linear differential operators, and are the infinitesimal generators of operators $e^{\theta X}$, $e^{\theta Y}$; we consider the product of $e^{\theta X}$ and $e^{\theta Y}$. In what follows we use the notation

$$[A,B] = AB - BA$$

for the *commutator bracket* of the operators A and B. A calculation shows that

$$e^{\theta X} e^{\theta Y} = e^{\theta X + \theta Y + (1/2)\theta^2[X,Y] + (1/12)\theta^3[X-Y,[X,Y]] + \cdots}.$$

Here, in the exponent on the right hand side, the $+ \cdots$ means terms of higher order in θ; it can be shown that each of these can be written using only commutator brackets $[-,-]$, without any products. If $[X,Y] = 0$, that is, if X and Y commute, then $e^{\theta X} e^{\theta Y} = e^{\theta(X+Y)}$, so that the composite $e^{\theta X} e^{\theta Y}$ of the two transformations coincides with the transformations $e^{\theta(X+Y)}$ corresponding to $X+Y$. In general these two do not coincide, but the difference between them can be computed by knowing the commutator bracket of the infinitesimal generators.

A *Lie algebra* is a vector space $\mathfrak{g}$, together with a law which associates to any two elements $X, Y \in \mathfrak{g}$ a bracket $[X,Y] \in \mathfrak{g}$, satisfying

(1) $\quad [X,Y] = -[Y,X]$,
(2) $\quad [[X,Y],Z] + [[Y,Z],X] + [[Z,X],Y] = 0, \qquad (1.13)$
(3) $\quad [\alpha X + \beta Y, Z] = \alpha[X,Z] + \beta[Y,Z]$.

(Here α, β are coefficients acting by scalar multiplication in the vector space $\mathfrak{g}$.) If we ignore worries about convergence, then for a Lie algebra $\mathfrak{g}$, the set of transformations

$$G = \{e^X \mid X \in \mathfrak{g}\}$$

is a group. To run ahead of ourselves, we note that when treating infinite dimensional symmetries, as we do in soliton theory, it often happens that the Lie algebra $\mathfrak{g}$ is comparatively easy to deal with, even in cases where handling the transformation group G might lead to difficulties.

1.2 Symmetries of the KdV equation

As explained above, there are two different contexts in which rotations are generated by an infinitesimal linear transformation:

in 2 dimensional space by $\begin{pmatrix} 0 & -1 \\ 1 & 0 \end{pmatrix}$;

in an infinite dimensional space by $x\dfrac{\partial}{\partial y} - y\dfrac{\partial}{\partial x}$.

We can also consider nonlinear infinitesimal transformations. For functions of two variables $u(x,t)$, consider the differential equation

$$\frac{\partial u}{\partial t} = u\frac{\partial u}{\partial x} + \frac{\partial^3 u}{\partial x^3}. \tag{1.14}$$

This is the KdV equation, the main theme of this chapter. Here we take the coefficients of $u(\partial u/\partial x)$ and $\partial^3 u/\partial x^3$ to be equal to 1, but we can make them into arbitrary nonzero constants by multiplying t, x and u by constant scaling factors. All of these are also called KdV equations. This equation describes an infinitesimal transformation in time t acting on a function u of x by the operator

$$K(u) = u\frac{\partial u}{\partial x} + \frac{\partial^3 u}{\partial x^3}. \tag{1.15}$$

Quite generally, an equation of the form $\partial u/\partial t = K(u)$ is called an *evolution equation*. The equation is said to be linear or nonlinear depending on the nature of the operator $K(u)$. If $K(u)$ is linear then $u \mapsto K(u)$ is just the infinitesimal generator described in Section 1.1.

From now on, we say that $K(u)$ is an infinitesimal generator also in the nonlinear case. We interpret the evolution equation, including the nonlinear case, as given by infinitesimal transformation of functions, and search for symmetries of the KdV equation among these. We pose the problem as follows: does the KdV equation

$$\frac{\partial u}{\partial t} = K(u) \tag{1.16}$$

have a symmetry of the form

$$\frac{\partial u}{\partial s} = \widehat{K}(u)? \tag{1.17}$$

What does it mean to say that (1.17) is a symmetry of (1.16)? Consider a function of three variables $u(x,t,s)$. In what follows, for simplicity, we write derivatives (and higher order derivatives) as

$$(\partial u/\partial t) = u_t, \quad \partial^3 u/\partial x^3 = u_{xxx} = u_{3x},$$

and so on. A polynomial in u and its x-derivatives u_x, u_{xx}, u_{3x}, ... is called a *differential polynomial* in u with respect to x. For example, (1.15) is a differential polynomial in u.

Let $\widehat{K}(u)$ be a differential polynomial in u. We consider (1.17) as an evolution equation in time s, and suppose that it can be solved with the given initial value $u(x,t,s=0)$. In other words, starting from the function of two variables $u(x,t,s=0)$ at time $s=0$, and solving (1.17), we get the function $u(x,t,s=\Delta s)$ at time Δs. To say that (1.17) gives a symmetry of the KdV equation means exactly that if $u(x,t,s=0)$ is a solution of (1.16) at time $s=0$, then so is $u(x,t,s)$ at any time s.

Treating t and s on an equal footing reinterprets the question as follows: suppose that t and s are two independent time variables, and that we are given a function $u(x,t=0,s=0)$ when both $t=0$ and $s=0$. Then there are two methods to determine the function $u(x,\Delta s, \Delta t)$ at time $(\Delta t, \Delta s)$, as shown in the following diagram:

$$
\begin{array}{ccc}
u(x,t=\Delta t, s=0) & \longrightarrow & u(x,t=\Delta t, s=\Delta s) \\
\uparrow & \overset{A}{\underset{B}{\longrightarrow}} & \uparrow \\
u(x,t=0,s=0) & \longrightarrow & u(x,t=0,s=\Delta s)
\end{array}
\tag{1.18}
$$

In this diagram, the up arrows stand for solving (1.16), and the right arrows for solving (1.17). The composite arrow A goes first up, then across; whereas B goes first across, then up. If A and B give the same result then (1.17) is clearly a symmetry of (1.16). Passing to the limiting case when Δt, Δs are very small in (1.18), we see that for A = B to

1.2 Symmetries of the KdV equation

hold, we must have

$$\frac{\partial}{\partial s}K(u) = \frac{\partial}{\partial t}\widehat{K}(u). \tag{1.19}$$

Now does (1.19) hold for an arbitrary choice of $\widehat{K}(u)$? For example, if we try setting $\widehat{K}(u) = u^2$, we get

$$\begin{aligned}\text{left-handside} &= (uu_x + u_{3x})_s = u^2 u_x + u(u^2)_x + (u^2)_{3x}\\ &= 3u^2 u_x + 6u_x u_{xx} + 2uu_{3x},\\ \text{right-handside} &= (u^2)_t = 2u^2 u_x + 2uu_{3x}.\end{aligned}$$

So we see that (1.19) fails without the additional condition $0 = u^2 u_x + 6u_x u_{xx}$. Thus for arbitrary choices of $\widehat{K}(u)$, (1.16) and (1.17) are not compatible, so that (1.17) is not a symmetry of (1.16).

If in (1.15) we give u degree 2, u_x degree 3 and u_{xx} degree 4, then the right-hand side is homogeneous of degree 5. Although we do not explain the reason behind it, in searching for symmetries, there is no loss of generality in our argument below in restricting the infinitesimal generating operator to be a homogeneous differential polynomial.

The general form of a homogeneous differential polynomial of degree 7 is

$$C_1 u^2 u_x + C_2 uu_{3x} + C_3 u_x u_{xx} + C_4 u_{5x}. \tag{1.20}$$

Substituting (1.20) for $\widehat{K}(u)$ in (1.17) and assuming $C_1 = 1$, we see, after a similar but lengthy calculation, that the coefficients C_i are uniquely determined by the condition that (1.16) and (1.17) are compatible. In fact, we get

$$\frac{\partial u}{\partial s} = u^2 u_x + 2uu_{3x} + 4u_x u_{xx} + \frac{6}{5}u_{5x}. \tag{1.21}$$

Remark 1.1 *There do exist homogeneous differential polynomials of degrees 3 (respectively 5), but the resulting symmetries of $u(x,t)$ correspond simply to the parallel translation $x \mapsto x + s$ (respectively $t \mapsto t + s$).*

If we set to work to calculate more systematically, we would find that there apparently exists just one symmetry in each odd degree. Now, how can we carry out the argument for every odd number? See Exercise 1.2 for an example other than the KdV equation.

1.3 The Lax form of an evolution equation (the approach via linear differential equations)

Consider the linear differential equation

$$Pw = k^2 w, \quad \text{where} \quad P = \frac{\partial^2}{\partial x^2} + u. \tag{1.22}$$

We think of u as given as a function of x, and P as an operator acting on functions of x. Thus k^2 is an *eigenvalue* of P, and k is called the *spectral* variable. If $u \equiv 0$ then $w = e^{kx}$ is one solution, but we can also look for solutions in the general case as formal power series of the form

$$w = e^{kx}\left(w_0 + \frac{w_1}{k} + \frac{w_2}{k^2} + \cdots\right). \tag{1.23}$$

Here *formal* means that we do not necessarily require the power series to converge. Substituting (1.23) in (1.22) gives

$$\frac{\partial w_0}{\partial x} = 0 \quad \text{and} \quad 2\frac{\partial w_j}{\partial x} + \frac{\partial^2 w_{j-1}}{\partial x^2} + u w_{j-1} = 0 \quad \text{for } j \geq 1.$$

Assuming that $w_0 \equiv 1$, the w_j can be determined successively (up to constants of integration) by integrating with respect to x.

We now introduce a time variable t, and allow the given function $u = u(x,t)$ to vary with t. We want to solve (1.22) in terms of a time evolution of w with a linear operator. The operator P in (1.22) is a second order differential operator, so this time we try looking for a third order differential operator

$$\frac{\partial w}{\partial t} = Bw, \quad \text{where} \quad B = \frac{\partial^3}{\partial x^3} + b_1 \frac{\partial}{\partial x} + b_2. \tag{1.24}$$

Solving this gives a function $w(x,t;k)$ of two variables x, t for any fixed value of k. We know that (for u independent of k) at time $t = 0$, the function $w(x, t = 0; k)$ satisfies (1.22). Does this continue to hold at other times t? (For this to make sense, $u = u(x,t)$ must also be independent of k.) If (1.22) holds then differentiating both sides with respect to t gives

$$\left(\frac{\partial P}{\partial t} + [P, B]\right) w = 0. \tag{1.25}$$

Here $[P, B] = PB - BP$ is the commutator bracket of the differential operators P and B, and $\partial P/\partial t = \partial u/\partial t$, where u is the given function. Thus (1.25) only involves derivatives with respect to x, so is an ordinary differential equation (independent of k). If (1.25) holds for an arbitrary

1.3 The Lax form of an evolution equation

value of the eigenvalue k then the ODE (1.25) has infinitely many independent solutions. This is impossible unless the differential equation is trivial. Thus we must have

$$\frac{\partial P}{\partial t} + [P, B] = 0. \tag{1.26}$$

Writing this out as conditions on the coefficients u of P and b_1, b_2 of B gives

$$\begin{aligned} b_1 &= \frac{3}{2}u, \\ b_2 &= \frac{3}{4}u_x, \\ \frac{\partial u}{\partial t} &= \frac{3}{2}uu_x + \frac{1}{4}u_{3x}. \end{aligned} \tag{1.27}$$

Here we are solving under the condition that u, b_1, b_2 and their x-derivatives tend to 0 as $x \to \pm\infty$. Thus (1.22) and (1.24) are compatible only if $u(x,t)$ is a solution of the KdV equation. The compatibility condition (1.26) is equivalent to the KdV equation, and is called the *Lax form* of the KdV equation.

We summarise the above argument schematically:

$$\boxed{\begin{array}{l} \text{linear system of equations:} \quad Pw = k^2 w \quad \text{and} \quad \dfrac{\partial w}{\partial t} = Bw \\[1ex] \qquad \Bigg\Downarrow \begin{array}{l} \text{compatibility} \\ \text{condition} \end{array} \\[2ex] \text{Lax representation of the KdV equation:} \quad \dfrac{\partial P}{\partial t} = [B, P]. \end{array}} \tag{1.28}$$

Replacing the third order linear differential operator B in x with linear differential operators of higher order gives rise to nonlinear evolution equations called the *higher order KdV equations*, which involve higher order derivatives. To see this clearly, we carry out a few algebraic preliminaries in the next chapter. See Exercise 1.3.

Exercises to Chapter 1

1.1. What is the function generated from $f(x) = x$ by the infinitesimal transformation $x^2 \partial/\partial x$?

1.2. Determine a symmetry of the equation

$$\frac{\partial u}{\partial t} = u^2 u_x + u_{xxx}.$$

[Hint: Set $\widehat{K}(u) = Au^4 u_x + Bu^2 u_{3x} + Cuu_x u_{2x} + Du_x^3 + Eu_{5x}$, calculate $(\partial/\partial t)(\widehat{K}(u))$ and $(\partial/\partial s)(K(u))$ by the method indicated after (1.19), then equate coefficients to determine A, B, C, D, E.]

1.3. What equation do you get from the Lax equation (1.26) if you swap the roles of P (1.22) and B (1.24)?

2
The KdV hierarchy

The value of mathematics is its unrestrained freedom of expression, the licence to introduce new concepts. You can probably still remember the amazing experience of meeting the complex numbers for the first time. In this chapter, we introduce the inverse of the differential operator $\partial/\partial x$. We then see the astonishing power with which this gives rise to the higher order KdV equations.

2.1 Pseudodifferential operators

In discussing operators, we usually have in mind some operations applied to functions. However, here we concentrate instead on the composition rules for operators. Depending on the situation, we may be able to dispense altogether with actual operations. As we see in what follows, by doing this, we can define negative powers of a differential operator.

For simplicity of notation, we write ∂ for the derivative with respect to x. For f a function of x, consider the operator $\partial^n \circ f$ obtained as the composite of multiplication by f with the differential operator ∂^n. We can write this out with all the differentials on the right, and $\partial^n \circ f$ becomes

$$\partial^n \circ f = \sum_{j \geq 0} \binom{n}{j} (\partial^j f) \circ \partial^{n-j} \qquad (2.1)$$

(compare Exercise 2.1). Here $\partial^j f$ is the function obtained as the jth derivative of f, and $\binom{n}{j}$ the binomial coefficient, which we think of as defined by the formula

$$\binom{n}{j} = \frac{n(n-1)\cdots(n-j+1)}{j(j-1)\cdots 1}. \qquad (2.2)$$

When j is a natural number, this definition makes sense for any value of n. Now if n is a positive integer then $\binom{n}{j} = 0$ whenever $j \geq n+1$, so that there is no harm in allowing the sum in (2.1) to run over all natural numbers j. Having done this, we can use (2.1) to define the composite of multiplication by f and the operator ∂^n for any value of n. More generally, we consider an expression of the form

$$L = \sum_{j=0}^{\infty} f_j \partial^{\alpha-j}, \qquad (2.3)$$

which we call a (formal) *pseudodifferential operator* of order $\leq \alpha$. The product of pseudodifferential operators is defined using (2.1); see Exercises 2.2–3.

Example 2.1 *We explain how to compute the square root of the Schrödinger operator $\partial^2 + u$. For this, set*

$$X = \partial + \sum_{n=1}^{\infty} f_n \partial^{-n} \qquad (2.4)$$

with unknown functions f_n, and compute the square:

$$X^2 = \partial^2 + 2 \sum_{n \geq 1} f_n \partial^{1-n} + \sum_{n \geq 1} (\partial f_n) \partial^{-n} + \sum_{\substack{m,n \geq 1 \\ l \geq 0}} \binom{-n}{l} f_n (\partial^l f_m) \partial^{-m-n-l}.$$

If we set $X^2 = \partial^2 + u$, we can solve term by term for the f_n. We write out the first few terms, obtaining

$$(\partial^2 + u)^{1/2} = \partial + \frac{1}{2} u \partial^{-1} - \frac{1}{4} u_x \partial^{-2} + \left(\frac{u_{xx}}{8} - \frac{u^2}{8} \right) \partial^{-3} + \cdots . \qquad (2.5)$$

We now explain how to define an action of the pseudodifferential operator L of (2.3) on a power series of the form

$$w = k^\beta e^{kx} \left(w_0 + \frac{w_1}{k} + \frac{w_2}{k^2} + \cdots \right). \qquad (2.6)$$

First of all, in the case $w_0 \equiv 1$, $w_1 \equiv w_2 \equiv \cdots \equiv 0$, it is natural to set

$$L(k^\beta e^{kx}) = k^{\alpha+\beta} e^{kx} \sum_{n=0}^{\infty} f_n k^{-n},$$

because $\partial^n(e^{kx}) = k^n e^{kx}$ for a natural number n. Now the power series w of (2.6) can be written in the form

$$w = M e^{kx}, \quad \text{where} \quad M = \sum_{j \geq 0} w_j \partial^{\beta-j}.$$

Thus the action of L on w can be defined by $Lw = L(Me^{kx}) = (L \circ M)(e^{kx})$. It is not quite obvious from the above argument that the action of a pseudodifferential operator is well defined and without contradiction, but we do not want to get involved in the proof at this point.

2.2 Higher order KdV equations

Let M be a pseudodifferential operator

$$M = \sum_{l=0}^{\infty} g_l \partial^{n-l} \tag{2.7}$$

of order $n \in \mathbb{Z}$. We define $M_\pm$ as follows:

$$M_+ = \sum_{l=0}^{n} g_l \partial^{n-l}, \quad \text{and} \quad M_- = M - M_+. \tag{2.8}$$

Thus M_+ is a differential operator.

Example 2.2 *We return to the Schrödinger operator $P = \partial^2 + u$ discussed in Example 2.1, and compute $(\partial^2 + u)^{3/2}$ using (2.5). Then comparing with (1.24) and (1.27), we see that*

$$B = \left((\partial^2 + u)^{3/2}\right)_+. \tag{2.9}$$

More generally, let l be a positive odd number. Set

$$B_l = \left((\partial^2 + u)^{l/2}\right)_+, \tag{2.10}$$

and consider the corresponding Lax form (1.26). We have $P = \partial^2 + u$ and $[P, P^{l/2}] = 0$, and we therefore get

$$[P, B_l] = -[P, (P^{l/2})_-]. \tag{2.11}$$

On the left-hand side, both P and B_l are differential operators, and hence so is $[P, B_l]$. On the right-hand side, P is a differential operator of order 2, and $(P^{l/2})_-$ a pseudodifferential operator of order ≤ -1, so that the order of $-[P, (P^{l/2})_-]$ as a pseudodifferential operator is at most $2 + (-1) - 1 = 0$ (see Exercise 2.4). Therefore (2.11) is a differential operator of degree 0, in other words, a function.

This is a differential polynomial in x. We write $K_l(u)$ for $[P, (P^{l/2})_-]$. On the other hand, $\partial P/\partial t$ is equal to $\partial u/\partial t$, and the Lax form (1.26) is equivalent to the nonlinear evolution equation in u

$$\frac{\partial u}{\partial t} = K_l(u).$$

14 *The KdV hierarchy*

When we derived (1.27) from (1.26), the computation of the coefficients b_1, b_2 was based on the condition that the terms of order greater than 1 in $[P, B]$ vanished, but these conditions are automatically satisfied by virtue of the treatment in (2.10).

2.3 Infinitely many commuting symmetries

We use variables x_l indexed by the positive odd integers l (compare Exercise 2.5) and consider the system of equations

$$\frac{\partial u}{\partial x_l} = K_l(u) \quad \text{for } l = 1, 3, 5, \ldots, \tag{2.12}$$

where $K_l(u) = -[P, (P^{l/2})_+]$.

In particular, for $i = 1, 3$, we get

$$\frac{\partial u}{\partial x_1} = u_x,$$

$$\text{and} \quad \frac{\partial u}{\partial x_3} = \frac{3}{2} u u_x + \frac{1}{4} u_{3x}.$$

In other words, $x_1 = x$, $x_3 = t$ (see Exercise 2.6). We now show that all of these are pairwise compatible. We must prove that

$$\frac{\partial}{\partial x_l} K_j(u) = \frac{\partial}{\partial x_j} K_l(u). \tag{2.13}$$

In particular, taking $j = 3$, we find that $K_l(u)$ give symmetries of the KdV equation. For general j, l, (2.13) asserts that the symmetries given by $K_j(u)$ and $K_l(u)$ commute. That is, composing the evolution in the x_j time direction and in the x_l time direction is independent of the order of composition, in the sense discussed in connection with (1.18).

Proof We have

$$\frac{\partial P}{\partial x_l} = -[P, (P^{l/2})_+].$$

Hence if $f(P)$ is any function of P, we have

$$\frac{\partial f(P)}{\partial x_l} = -[f(P), (P^{l/2})_+].$$

Therefore

$$\frac{\partial}{\partial x_l}(P^{j/2})_+ = \left(\frac{\partial}{\partial x_l} P^{j/2}\right)_+ = -\left([P^{j/2}, (P^{l/2})_+]\right)_+,$$

2.3 Infinitely many commuting symmetries

so that

$$\frac{\partial}{\partial x_l} K_j(u) = -\frac{\partial}{\partial x_l}[P, (P^{j/2})_+]$$

$$= \left[[P, (P^{l/2})_+], (P^{j/2})_+\right] + \left[P, \left([P^{j/2}, (P^{l/2})_+]\right)_+\right]. \quad (2.14)$$

Now using the fact that

$$\left([P^{j/2}, (P^{l/2})_+]\right)_+ = \left[(P^{j/2})_+, (P^{l/2})_+\right] + \left([(P^{j/2})_-, (P^{l/2})_+]\right)_+$$

$$= \left[(P^{j/2})_+, (P^{l/2})_+\right] - \left([(P^{j/2})_+, P^{l/2}]\right)_+,$$

the two terms on the right-hand side of (2.14) are

$$\left[P, \left[(P^{j/2})_+, (P^{l/2})_+\right]\right] - \left[P, \left([(P^{j/2})_+, P^{l/2}]\right)_+\right]. \quad (2.15)$$

Thus applying the Jacobi identity (1.13) to (2.14) and (2.15), the left-hand side of (2.13) becomes

$$\left[P, \left([P^{l/2}, (P^{j/2})_+]\right)_+\right] + \left[[P, (P^{j/2})_+], (P^{l/2})_+\right],$$

and this is exactly the far right side (2.14) with j and l interchanged.

Q.E.D.

The equation (2.12) obtained in this way is the lth order KdV equation. The whole system of equations is called the *KdV hierarchy*. We summarise our result as follows:

linear system of equations:	$\begin{cases} Pw = k^2 w \\ \dfrac{\partial w}{\partial x_l} = (L^l)_+ w, \end{cases}$ where $L^2 = P$.
⇓ compatibility conditions	
KdV hierarchy:	$\dfrac{\partial P}{\partial x_l} = [(L^l)_+, P].$

That is, by searching for symmetries of the KdV equation as infinitesimal transformations of the form (2.12), we have obtained infinitely many pairwise commuting symmetries. In fact, these are not the only symmetries of the KdV equation. There is much bigger noncommuting symmetry. Before going into this, we generalise our considerations so far.

2.4 The KP hierarchy

In place of the square root of the Schrödinger operator, $(\partial^2 + u)^{1/2}$, we consider a pseudodifferential operator,

$$L = \partial + \sum_{j=1}^{\infty} f_j \partial^{-j} \qquad (2.16)$$

of order 1, and the corresponding eigenvalue problem:

$$Lw = kw. \qquad (2.17)$$

We prepare an infinite set of variables $\mathbf{x} = (x_1, x_2, x_3, \ldots)$, and identify $x_1 = x$. Now consider a formal solution

$$w = e^{\xi(\mathbf{x},k)} \left(1 + \frac{w_1}{k} + \frac{w_2}{k^2} + \cdots \right), \quad \text{where} \quad \xi(\mathbf{x},k) = \sum_{j=1}^{\infty} x_j k^j \qquad (2.18)$$

(see Exercise 2.7). Note here that

$$\frac{\partial}{\partial x_j} e^{\xi(\mathbf{x},k)} = k^j e^{\xi(\mathbf{x},k)}. \qquad (2.19)$$

Considering the linear system of equations

$$\frac{\partial w}{\partial x_j} = B_j w, \quad \text{where} \quad B_j = (L^j)_+, \qquad (2.20)$$

we find that the compatibility condition between (2.17) and (2.20) is

$$\frac{\partial L}{\partial x_j} = [B_j, L]. \qquad (2.21)$$

This is an infinite set of nonlinear evolution equations in infinitely many functions $f_1, f_2, \ldots$ of the infinitely many variables $(x_1, x_2, x_3, \ldots)$. Equation (2.21) is called the *KP hierarchy*.

If L satisfies the KdV condition

$$(L^2)_- = 0$$

then we get back the KdV hierarchy. In this case $L^2 = \partial^2 + u$, and all those infinitely many functions $f_1, f_2, \ldots$ are determined by the single function u. Also, if j is even, we have $[B_j, L] = 0$, and so

$$\frac{\partial u}{\partial x_j} = 0.$$

Thus of the infinitely many variables, only those with odd index are meaningful.

2.4 The KP hierarchy

As defined above, the KP hierarchy is an equation in infinitely many unknown functions, but it can be reduced to a single unknown function. This unknown function τ is called a *tau function*. In what follows we discuss this without proofs.

We consider the problem of looking for a formal solution of the linear equation (2.20) in the form

$$w = M e^{\xi(\mathbf{x},k)}, \quad \text{where} \quad M = 1 + \sum_{j=1}^{\infty} w_j \partial^{-j}. \tag{2.22}$$

Substituting (2.22) in (2.17) gives the relation

$$L = M \circ \partial \circ M^{-1} \tag{2.23}$$

between pseudodifferential operators (see Exercise 2.8). By (2.23), we can use $(w_1, w_2, \ldots)$ in place of the unknown functions $(f_1, f_2, \ldots)$. In fact, by the compatibility condition (2.21), it turns out that these can all be written in terms of a single function τ:

$$w = \frac{\tau\left(x_1 - \frac{1}{k}, x_2 - \frac{1}{2k^2}, x_3 - \frac{1}{3k^3}, \ldots\right)}{\tau(x_1, x_2, x_3, \ldots)} e^{\xi(\mathbf{x},k)} \tag{2.24}$$

(see Exercise 2.9). From this, we can determine $(w_1, w_2, \ldots)$ in terms of τ: for example,

$$w_1 = -\frac{\partial \tau}{\partial x_1} \bigg/ \tau, \tag{2.25}$$

$$\text{and} \quad w_2 = \frac{1}{2}\left(\frac{\partial^2 \tau}{\partial x_1^2} - \frac{\partial \tau}{\partial x_2}\right) \bigg/ \tau. \tag{2.26}$$

Thus the KP hierarchy can be viewed as an infinite set of nonlinear differential equations in a function τ of infinitely many variables $(x_1, x_2, x_3, \ldots)$. In what follows, the tau function τ plays a fundamental role in discussing the noncommuting symmetries of the KdV and KP hierarchies. The original unknown function u of the KdV equation can be written in terms of τ as follows:

$$u = 2\frac{\partial^2}{\partial x^2} \log \tau. \tag{2.27}$$

The KP hierarchy includes the following equation in u (see Exercise 2.10)

$$\frac{3}{4}\frac{\partial^2 u}{\partial x_2^2} = \frac{\partial}{\partial x}\left(\frac{\partial u}{\partial x_3} - \frac{3}{2}u\frac{\partial u}{\partial x} - \frac{1}{4}\frac{\partial^3 u}{\partial x^3}\right). \tag{2.28}$$

Exercises to Chapter 2

2.1. Explain the relation between equation (2.1) and the Leibniz rule of differential calculus.

2.2. If $L = \sum_{k=0}^{\infty} f_k \partial^{\alpha-k}$ and $M = \sum_{k=0}^{\infty} g_k \partial^{\beta-k}$, compute the composite $L \circ M$ (also abbreviated to LM).

2.3. Compute $(\partial + x)^{-1}$.

2.4. Let L_1 and L_2 be pseudodifferential operators of orders α_1 and α_2; what are the orders of $L_1 L_2$ and $[L_1, L_2]$?

2.5. Explain why there is no point in considering even values of l in (2.12).

2.6. What is $\dfrac{\partial u}{\partial x_5}$ in (2.12)?

2.7. Derive the relations holding between f_1, f_2 in (2.16) and w_1, w_2 in (2.18).

2.8. Determine $M^{-1} = 1 + v_1 \partial^{-1} + v_2 \partial^{-2} + \cdots$.

2.9. Equations (2.25) and (2.26) can be rewritten

$$\frac{\partial \log \tau}{\partial x_1} = -w_1,$$

$$\frac{\partial \log \tau}{\partial x_2} = -2w_2 + w_1^2 - \frac{\partial w_1}{\partial x_1}.$$

Prove that these two relations are compatible.

2.10. Derive (2.28).

3
The Hirota equation and vertex operators

Hirota's theory of equations of bilinear type is a classic instance of freedom of expression in mathematics. In the 1970s, Hirota introduced an effective method for constructing solutions of the KdV equation and other soliton equations, although at the time it was not clear that his methods had any connection with other areas of mathematics. However, a useful idea in mathematics does not remain in isolation for long. We will see how the Hirota equation relates to the vertex operators from elementary particle theory.

3.1 The Hirota derivative

Given two functions $f(x)$ and $g(x)$ of a single variable x, we can write out the Taylor expansion of $f(x+y)g(x-y)$ around $y = 0$ in the form

$$f(x+y)g(x-y) = \sum_{j=0}^{\infty} \frac{1}{j!} (\mathrm{D}_x^j f \cdot g) y^j. \tag{3.1}$$

The operator $(f, g) \mapsto \mathrm{D}_x^j f \cdot g$ is the *Hirota derivative*.

Example 3.1 *We have*

$$\mathrm{D}_x f \cdot g = \frac{\partial f}{\partial x} g - f \frac{\partial g}{\partial x},$$

$$\mathrm{D}_x^2 f \cdot g = \frac{\partial^2 f}{\partial x^2} g - 2 \frac{\partial f}{\partial x} \frac{\partial g}{\partial x} + f \frac{\partial^2 g}{\partial x^2}.$$

Note that $\mathrm{D}_x^j f \cdot g$ is a single entity, and the D_x are not to be thought of as some kind of individual operators. Thus $\mathrm{D}_x^j f \cdot g$ is definitely not an object called $\mathrm{D}_x^j f$ multiplying g. The Hirota derivative in many variables is defined in the same way. That is, if $f = f(x_1, x_2, \ldots)$ and $g = g(x_1, x_2, \ldots)$, we have

$$e^{y_1 D_1 + y_2 D_2 + \cdots} f \cdot g = f(x_1 + y_1, x_2 + y_2, \ldots) g(x_1 - y_1, x_2 - y_2, \ldots),$$

where the translation of the coordinates $x_i \mapsto x_i \pm y_i$ is as in (1.12). Expanding the left-hand side as a Taylor series in $(y_1, y_2, \ldots)$, we get

$$f \cdot g + y_1 (D_1 f \cdot g) + y_2 (D_2 f \cdot g) + \cdots + \frac{1}{2} y_1^2 (D_1^2 f \cdot g) + \cdots,$$

and comparing with the right-hand side defines all the Hirota derivatives.

Example 3.2 *Suppose that f is a function of two variables (x, t). Then*

$$D_t D_x f \cdot f = 2 \left(\frac{\partial^2 f}{\partial t \partial x} f - \frac{\partial f}{\partial t} \frac{\partial f}{\partial x} \right).$$

The following equations hold:

$$\frac{\partial^2}{\partial x^2} \log f = \frac{1}{2f^2} (D_x^2 f \cdot f);$$

$$\frac{\partial^4}{\partial x^4} \log f = \frac{1}{2f^2} (D_x^4 f \cdot f) - 6 \left(\frac{1}{2f^2} (D_x^2 f \cdot f) \right)^2.$$

Now we view (2.27) as defining a new unknown function τ, and rewrite the KdV equation (1.27) as an equation in τ. Then after carrying out one integration with respect to x, we get

$$8 \frac{\partial^2}{\partial t \partial x} \log \tau = 3 \left(2 \frac{\partial^2}{\partial x^2} \log \tau \right)^2 + 2 \frac{\partial^4}{\partial x^4} \log \tau.$$

Using the above formulas, we can rewrite this as follows in terms of Hirota derivatives:

$$(4 D_t D_x - D_x^4) \tau \cdot \tau = 0. \tag{3.2}$$

We write $(D_1, D_2, \ldots)$ for the Hirota derivatives with respect to the variables $(x_1, x_2, \ldots)$. A *Hirota equation* is an equation of the form

$$P(D_1, D_2, \ldots) \tau \cdot \tau = 0, \tag{3.3}$$

where $P(D_1, D_2, \ldots)$ is a polynomial in $(D_1, D_2, \ldots)$. Consider how we might solve it. If P is an odd function then $P\tau \cdot \tau$ is trivially 0, independently of τ. For example, $D_x \tau \cdot \tau = (\partial \tau / \partial x) \tau - \tau (\partial \tau / \partial x) = 0$. Thus we suppose that P is an even function, that is,

$$P(D_1, D_2, \ldots) = P(-D_1, -D_2, \ldots).$$

3.1 The Hirota derivative

Suppose also that $P(0) = 0$. First, $\tau \equiv 1$ is always a solution. Thus we look for a solution by expanding it as

$$\tau = 1 + \varepsilon f_1 + O(\varepsilon^2).$$

Taking the degree 1 term in ε of (3.3) gives a linear equation in f_1:

$$P(\partial_1, \partial_2, \ldots) f_1 = 0, \quad \text{where } \partial_j = \frac{\partial}{\partial x_j}. \tag{3.4}$$

If we choose a set of complex numbers $k_1, k_2, \ldots$ satisfying

$$P(k_1, k_2, \ldots) = 0$$

then

$$f_1 = e^{k_1 x_1 + k_2 x_2 + \cdots}$$

satisfies (3.4). Or more generally, if we prepare several such sets $(k_1^{(j)}, k_2^{(j)}, \ldots)$ with $P(k_1^{(j)}, k_2^{(j)}, \ldots) = 0$ then

$$f_1 = \sum_{j=1}^{n} c_j e^{k_1^{(j)} x_1 + k_2^{(j)} x_2 + \cdots} \tag{3.5}$$

is again a solution. In general, $1 + \varepsilon f_1$ is not a solution of (3.3) to all orders. However, in the case $n = 1$, we can break off the expansion at the linear term in ε, and get that

$$\tau = 1 + \varepsilon e^{k_1 x_1 + k_2 x_2 + \cdots}$$

is a solution of (3.3), as can be verified by a direct calculation. For example, for the KdV equation, writing c for ε, we get

$$\tau = 1 + c e^{2kx + 2k^3 t}$$

as a solution of (3.2). A solution obtained in this way is a 1-soliton solution.

More generally, in the Hirota equation, a solution that can be expressed as a polynomial in exponentials

$$e^{k_1 x_1 + k_2 x_2 + \cdots}$$

of the variables $x_1, x_2, \ldots$ is called a *soliton solution*. Here

$$k_1 x_1 + k_2 x_2 + \cdots$$

is called an *exponent*. In particular, a soliton solution having n distinct exponents is called an *n-soliton solution*.

The special feature of the KdV equation as an integrable system is not just that it can be rewritten as a Hirota equation, but that for arbitrary n, there exists a solution having linear approximation $1 + \varepsilon f_1$, with f_1 as in (3.5); that is, there exists an n-soliton solution. A general Hirota equation always has soliton solutions for $n = 1, 2$. As an empirical fact, the existence of n-soliton solutions for $n \geq 3$ is more or less always equivalent to the integrability of the system.

Let us consider $n = 2$. Suppose that $(k_1^{(j)}, k_2^{(j)}, \ldots)$ for $j = 1, 2$ are given, satisfying $P(k_1^{(j)}, k_2^{(j)}, \ldots) = 0$. We consider

$$\tau = 1 + \varepsilon \sum_{j=1}^{2} c_j e^{k_1^{(j)} x_1 + k_2^{(j)} x_2 + \cdots} + \varepsilon^2 f_2 + O(\varepsilon^3).$$

Taking the second order terms in ε in the equation $P(\mathrm{D}_1, \mathrm{D}_2, \ldots)\tau \cdot \tau = 0$ gives

$$P(\partial_1, \partial_2, \ldots) f_2$$
$$+ c_1 c_2 P(k_1^{(1)} - k_1^{(2)}, k_2^{(1)} - k_2^{(2)}, \ldots) e^{(k_1^{(1)} + k_1^{(2)}) x_1 + (k_2^{(1)} + k_2^{(2)}) x_2 + \cdots} = 0.$$

Solving this gives

$$f_2 = -\frac{P(k_1^{(1)} - k_1^{(2)}, k_2^{(1)} - k_2^{(2)}, \ldots)}{P(k_1^{(1)} + k_1^{(2)}, k_2^{(1)} + k_2^{(2)}, \ldots)} c_1 c_2 e^{(k_1^{(1)} + k_1^{(2)}) x_1 + (k_2^{(1)} + k_2^{(2)}) x_2 + \cdots}$$

and the expression truncated at ε^2 is a 2-soliton solution. For $P = 4\mathrm{D}_t \mathrm{D}_x - \mathrm{D}_x^4$, setting $k^{(1)} = (2k_1, 2k_1^3)$ and $k^{(2)} = (2k_2, 2k_2^3)$ gives

$$f_2(x, t) = \frac{(k_1 - k_2)^2}{(k_1 + k_2)^2} c_1 c_2 e^{2(k_1 + k_2) x + 2(k_1^3 + k_2^3) t}$$

(see Exercise 3.1).

3.2 n-Solitons

We write down a formula for an n-soliton solution of the KdV equation. In Section 3.4 we prove that this is the tau function of the KdV equation. We prepare parameters $c_1, \ldots, c_n$ and $k_1, \ldots, k_n$, and extend the two variables (x, t) to an infinite number of variables $x_1 = x$, $x_3 = t$, $x_5, x_7, \ldots$. For the infinitely many variables $x_1, x_2, x_3, \ldots$ of (2.18), we introduce the notation

$$\xi(\mathbf{x}, k) = \sum_{i=1}^{\infty} x_i k^i,$$

and define the exponents ξ_i and factors $a_{ii'}$ by

$$\xi_i = 2\sum_{j=0}^{\infty} k_i^{2j+1} x_{2j+1} = \xi(\mathbf{x}, k_i) - \xi(\mathbf{x}, -k_i), \tag{3.6}$$

$$a_{ii'} = \frac{(k_i - k_{i'})^2}{(k_i + k_{i'})^2}. \tag{3.7}$$

We set $I = \{1, \ldots, n\}$. The sum over all subsets J of I

$$\tau(x_1, x_3, \ldots) = \sum_{J \subset I} \left(\prod_{i \in J} c_i\right) \left(\prod_{\substack{i, i' \in J \\ i < i'}} a_{ii'}\right) \exp\left(\sum_{i \in J} \xi_i\right) \tag{3.8}$$

is the n-soliton.

Example 3.3 *In the case $n = 3$,*

$$\tau = 1 + c_1 e^{\xi_1} + c_2 e^{\xi_2} + c_3 e^{\xi_3}$$
$$+ c_1 c_2 a_{12} e^{\xi_1 + \xi_2} + c_1 c_3 a_{13} e^{\xi_1 + \xi_3} + c_2 c_3 a_{23} e^{\xi_2 + \xi_3}$$
$$+ c_1 c_2 c_3 a_{12} a_{13} a_{23} e^{\xi_1 + \xi_2 + \xi_3}.$$

In other words, in the above notation, $\tau(x_1, x_3, \ldots)$ satisfies

$$(4D_1 D_3 - D_1^4)\tau \cdot \tau = 0$$

(see Exercise 3.2).

We add a few words concerning the fact that we have added infinitely many variables. These variables correspond to the commuting symmetries of the KdV equation considered in Chapter 1, that is, to the KdV hierarchy. Let us consider the problem of whether the higher order KdV equations can be written in Hirota form. If we write the order $2j + 1$ KdV equation in the usual way in terms of the unknown function u, it can be written as an equation in the two variables x_1 and x_{2j+1}.

However, if we try to rewrite it directly in Hirota form using D_1 and D_{2j+1}, it just does not work out properly. To put things differently, for any n-soliton (3.8), we can try to reformulate the problem as trying to find a polynomial P satisfying

$$P(D_1, D_3, \ldots)\tau \cdot \tau = 0. \tag{3.9}$$

This is not a Hirota equation involving only D_1 and D_{2j+1}, so that we are looking for an equation containing an arbitrary finite number of Hirota derivatives. If the solution is (3.8), looking for an equation of the form (3.9), we find, for example,

$$D_1^6 - 20D_1^3 D_3 - 80D_3^2 + 144D_1 D_5.$$

Studying this in more detail, and counting D_{2j+1} as order $(2j+1)$, we find the following numbers of equations of order l (including trivial ones):

order	number of equations
1	1
2	0
3	2
4	1
5	3
6	2
7	5

More generally, the number of KdV equations of order m of Hirota type equals $A - B$, where A and B are the numbers of ways of writing m as a sum of positive integers, with the following restrictions:

$A = \#\{\text{partitions of } m \text{ as a sum of odd positive integers}\},$

$B = \#\{\text{partitions of } m \text{ as a sum of positive integers} \equiv 2 \mod 4\}.$

For example, there is one further linearly independent equation in order 6, namely $D_1^6 + 4D_1^3 - 32D_3^2$. One can ask if the equations arising in this way, when translated back into equations for the function u, really coincide with the KdV hierarchy considered in Chapter 1. This is in fact the case, but we do not go further into this question in this book. (However, compare (3.27).)

3.3 Vertex operators

The introduction of infinitely many variables was a crucial step for the purpose of displaying the symmetries of the KdV equation. This is clear, first of all, because these variables are just another way of expressing the existence of the commuting symmetries. Second, and even more necessary, they play a fundamental role in describing the noncommuting symmetries which we treat from now on.

We introduced the idea of infinitesimal transformations in (1.10). We use ε in place of θ, and consider an infinitesimal transformation of the tau function given by

$$\frac{\partial}{\partial \varepsilon}\tau(x_1, x_3, \ldots) = X\tau(x_1, x_3, \ldots).$$

We want to search for infinitesimal transformations of this type that transform a solution τ of the KdV hierarchy into another solution. In fact, we will see that, starting from the n-soliton solution τ_n (3.8), there exists an X such that

$$\tau_{n+1} = e^{\varepsilon X}\tau_n$$

is an $(n+1)$-soliton solution.

Let k be a parameter, and consider the linear operator

$$X(k) = \exp\left(2\sum_{j=0}^{\infty} k^{2j+1} x_{2j+1}\right) \exp\left(-2\sum_{j=0}^{\infty} \frac{1}{(2j+1)k^{2j+1}} \frac{\partial}{\partial x_{2j+1}}\right). \tag{3.10}$$

An operator of this form is called a *vertex operator*. The name comes from the theory of elementary particles, but we do not go into this. The action of $X(k)$ takes a function $f(x_1, x_3, \ldots)$ into

$$X(k)f(x_1, x_3, \ldots) = \exp\left(2\sum_{j=0}^{\infty} k^{2j+1} x_{2j+1}\right) f\left(x_1 - \frac{2}{k}, x_3 - \frac{2}{3k^3}, \ldots\right),$$

where the translation of the coordinates $x_1 \mapsto x_1 - 2/k$, etc., is interpreted as in (1.12).

Lemma 3.1

$$X(k_1)X(k_2) = \frac{(k_1 - k_2)^2}{(k_1 + k_2)^2} \exp\left(2\sum_{i=1}^{2}\sum_{j=0}^{\infty} k_i^{2j+1} x_{2j+1}\right)$$

$$\times \exp\left(-2\sum_{i=1}^{2}\sum_{j=0}^{\infty} \frac{1}{(2j+1)k_i^{2j+1}} \frac{\partial}{\partial x_{2j+1}}\right).$$

Proof We want to prove that the two operators A and B given by

$$A = -2\sum_{j=0}^{\infty} \frac{1}{(2j+1)k_1^{2j+1}} \frac{\partial}{\partial x_{2j+1}},$$

$$B = 2\sum_{j=0}^{\infty} k_2^{2j+1} x_{2j+1}$$

have exponentials e^A, e^B satisfying

$$e^A e^B = \frac{(k_1-k_2)^2}{(k_1+k_2)^2} e^B e^A.$$

We use the following formula, leaving the proof as an exercise (see Exercise 3.3):

$$[A,B] \text{ a scalar} \implies e^A e^B e^{-A} = e^{[A,B]} e^B. \tag{3.11}$$

(Here by a scalar, we mean an operator not involving either differentiation with respect to the variables $x_1, x_2, \ldots$, or multiplication.) In our case,

$$[A,B] = -4\sum_{j=0}^{\infty} \frac{1}{2j+1} \left(\frac{k_2}{k_1}\right)^{2j+1}$$

$$= 2\sum_{l=1}^{\infty} \frac{1}{l}\left(-\frac{k_2}{k_1}\right)^l - 2\sum_{l=1}^{\infty} \frac{1}{l}\left(\frac{k_2}{k_1}\right)^l$$

$$= -\log\left(1+\frac{k_2}{k_1}\right)^2 + \log\left(1-\frac{k_2}{k_1}\right)^2,$$

so that $[A,B]$ is a scalar,

$$e^{[A,B]} = \frac{(k_1-k_2)^2}{(k_1+k_2)^2},$$

and the lemma follows. Q.E.D.

As we see from the proof, it would be more accurate to describe the expression $(k_1-k_2)^2/(k_1+k_2)^2$ in the lemma as its Taylor expansion with respect to k_2/k_1. From the lemma, we get in particular

$$e^{cX(k)} = 1 + cX(k).$$

Therefore $e^{cX(k)}1$ gives a 1-soliton solution. In general the n-soliton solution in the form (3.8) is clearly obtained by taking

$$\tau = e^{c_1 X(k_1)} \cdots e^{c_n X(k_n)} 1. \tag{3.12}$$

3.3 Vertex operators

Let us write out the vertex operator and the n-soliton solution for the KP equation. Writing the KP equation (2.28) in Hirota form gives

$$(D_1^4 + 3D_2^2 - 4D_1D_3)\tau \cdot \tau = 0. \tag{3.13}$$

We leave the computation as an interesting exercise. Set $P(k_1, k_2, k_3) = k_1^4 + 3k_2^2 - 4k_1k_3$; then the solutions of $P(k_1, k_2, k_3) = 0$ are

$$(k_1, k_2, k_3) = (p - q, p^2 - q^2, p^3 - q^3) \tag{3.14}$$

for any p, q. Moreover, given two solutions

$$(k_1, k_2, k_3) = (p_1 - q_1, p_1^2 - q_1^2, p_1^3 - q_1^3),$$
$$\text{and} \quad (k_1', k_2', k_3') = (p_2 - q_2, p_2^2 - q_2^2, p_2^3 - q_2^3),$$

we get

$$-\frac{P(k_1 - k_1', k_2 - k_2', k_3 - k_3')}{P(k_1 + k_1', k_2 + k_2', k_3 + k_3')} = \frac{(p_1 - p_2)(q_1 - q_2)}{(p_1 - q_2)(q_1 - p_2)}.$$

We can set

$$\xi_i = \sum_{j=1}^{\infty}(p_i^j - q_i^j)x_j = \xi(\mathbf{x}, p_i) - \xi(\mathbf{x}, q_i), \tag{3.15}$$

$$a_{ii'} = \frac{(p_i - p_{i'})(q_i - q_{i'})}{(p_i - q_{i'})(q_i - p_{i'})}. \tag{3.16}$$

in the 1-soliton solution (3.8). The vertex operator $X(p,q)$ that gives rise to the n-soliton solution is

$$X(p,q) = \exp\left(\sum_{j=1}^{\infty}(p^j - q^j)x_j\right)\exp\left(-\sum_{j=1}^{\infty}\frac{1}{j}(p^{-j} - q^{-j})\frac{\partial}{\partial x_j}\right) \tag{3.17}$$

(see Exercise 3.4). Then

$$\tau = e^{c_1 X(p_1, q_1)} \cdots e^{c_n X(p_n, q_n)} 1 \tag{3.18}$$

is an n-soliton solution. To write this more concretely, in view of (3.15) and (3.16), it becomes

$$\tau(x_1, x_2, \ldots) = \sum_{J \subset I}\left(\prod_{i \in J} c_i\right)\left(\prod_{\substack{i,i' \in J \\ i < i'}} a_{ii'}\right)\exp\left(\sum_{i \in J}\xi_i\right). \tag{3.19}$$

In the following section we prove that this tau function actually satisfies the KP hierarchy. Specialising by setting $p_i = -q_i$ reduces (3.19) to (3.8), and so we see at the same time that (3.8) satisfies the KdV hierarchy.

3.4 The bilinear identity

We now prove that the tau function (3.18) is actually a solution of the KP hierarchy. The key to this is the following result.

Theorem 3.2 (Bilinear identity) *For arbitrary* $\mathbf{x}$ *and* $\mathbf{x}'$, *set*

$$\xi = \xi(\mathbf{x}, k), \quad \xi' = \xi(\mathbf{x}', k).$$

Then the following identity holds:

$$0 = \oint \frac{dk}{2\pi i} e^{\xi - \xi'} \tau\left(x_1 - \frac{1}{k}, x_2 - \frac{1}{2k^2}, \ldots\right) \tau\left(x'_1 + \frac{1}{k}, x'_2 + \frac{1}{2k^2}, \ldots\right). \tag{3.20}$$

Here the contour integration $\oint \frac{dk}{2\pi i}$ means that we expand the integrand around $k = \infty$ and take the coefficient of k^{-1}. If the integrand is a holomorphic function on the complex plane except for poles at a finite number of points $k \in \mathbb{C}$, this can be computed as the sum of residues at these poles (see Exercise 3.5).

Proof We prove (3.20). We have to show that if in (3.19) we change x_j to $x_j - 1/jk^j$ and compute the pole in k, the sum of the residues is 0. Computing the value of the ξ_i of (3.15) at $(x_1 - 1/k, x_2 - 1/2k^2, \ldots)$ gives

$$\exp\left(\sum_{j=1}^{\infty}(p_i^j - q_i^j)\left(x_j - \frac{1}{jk^j}\right)\right) = \frac{k - p_i}{k - q_i} e^{\xi_i}.$$

In the same way,

$$\exp\left(\sum_{j=1}^{\infty}(p_i^j - q_i^j)\left(x'_j - \frac{1}{jk^j}\right)\right) = \frac{k - q_i}{k - p_i} e^{\xi'_i}.$$

The residue at $k = q_i$ is

$$\sum_{i \in J \subset I}\left(\prod_{l \in J} c_l\right)\left(\prod_{\substack{l, l' \in J \\ l < l'}} a_{ll'}\right) \exp\left(\sum_{l \in J \setminus \{i\}} \xi_l\right)(q_i - p_i) \prod_{l \in J \setminus \{i\}} \frac{q_i - p_l}{q_i - q_l} e^{\xi(\mathbf{x}, p_i)}$$

$$\times \sum_{i \notin J' \subset I}\left(\prod_{l \in J'} c_l\right)\left(\prod_{\substack{l, l' \in J' \\ l < l'}} a_{ll'}\right) \exp\left(\sum_{l \in J'} \xi'_l\right) \frac{q_i - q_l}{q_i - p_l} e^{-\xi(\mathbf{x}', q_i)}$$

3.4 The bilinear identity

$$= c_i(q_i - p_i)e^{\xi(\mathbf{x},p_i)-\xi(\mathbf{x}',q_i)} \sum_{i \notin J \subset I} \left(\prod_{l \in J} c_l\right) \left(\prod_{\substack{l,l' \in J \\ l < l'}} a_{ll'}\right) \prod_{l \in J} \frac{p_i - p_l}{p_i - q_l} e^{\sum_{l \in J} \xi_l}$$

$$\times \sum_{i \notin J' \subset I} \left(\prod_{l \in J'} c_l\right) \left(\prod_{\substack{l,l' \in J' \\ l < l'}} a_{ll'}\right) \prod_{l \in J'} \frac{q_i - q_l}{q_i - p_l} e^{\sum_{l \in J'} \xi'_l}.$$

The residue at $k = p_i$ can be computed in the same way, and the two cancel exactly. Q.E.D.

We can use the bilinear identity to deduce that the tau function provides solutions to the KP hierarchy in each of the two following senses:

(i) the linear system of equations (2.20), and
(ii) the bilinear system of equations (3.9).

We set

$$w(\mathbf{x},k) = \frac{\tau(x_1 - \frac{1}{k}, x_2 - \frac{1}{2k^2}, \ldots)}{\tau(x_1, x_2, \ldots)} e^{\xi(\mathbf{x},k)}, \quad (3.21)$$

$$w^*(\mathbf{x},k) = \frac{\tau(x_1 + \frac{1}{k}, x_2 + \frac{1}{2k^2}, \ldots)}{\tau(x_1, x_2, \ldots)} e^{-\xi(\mathbf{x},k)}. \quad (3.22)$$

Then these are of the form

$$w(\mathbf{x},k) = e^{\xi(\mathbf{x},k)} \left(1 + \sum_{l=1}^{\infty} \frac{w_l}{k^l}\right), \quad (3.23)$$

$$w^*(\mathbf{x},k) = e^{-\xi(\mathbf{x},k)} \left(1 + \sum_{l=1}^{\infty} \frac{w_l^*}{k^l}\right), \quad (3.24)$$

but as we have just proved, these satisfy

$$\oint \frac{dk}{2\pi i} w(\mathbf{x},k) w^*(\mathbf{x}',k) = 0. \quad (3.25)$$

Let us derive (2.20) from (3.25). First we make two general observations concerning (3.25).

(1) If Q is any differential operator in variables $(x_1, x_2, \ldots)$ then

$$\oint \frac{dk}{2\pi i} (Qw(\mathbf{x},k)) w^*(\mathbf{x}',k) = 0.$$

(2) If a power series of the form

$$\widetilde{w}(\mathbf{x}, k) = e^{\xi(\mathbf{x},k)} \cdot \sum_{l=1}^{\infty} \frac{\widetilde{w}_l}{k^l} \qquad (3.26)$$

satisfies

$$\oint \frac{\mathrm{d}k}{2\pi \mathrm{i}} \widetilde{w}(\mathbf{x}, k) w^*(\mathbf{x}', k) = 0$$

then $\widetilde{w}_1 \equiv \widetilde{w}_2 \equiv \cdots \equiv 0$. (See Exercise 3.6).

We prove (2.20). Define L by (2.22) and (2.23), with the w we are considering. Then setting

$$Q = \frac{\partial}{\partial x_j} - (L^j)_+$$

gives

$$Qw = \frac{\partial w}{\partial x_j} - L^j w + (L^j)_- w,$$

so that by (2.17), $(\partial w/\partial x_j) - L^j w$ is of the form (3.26). The same holds for $(L^j)_- w = (L^j)_M \exp(\sum_{j=1}^{\infty} k^j x_j)$. Therefore Qw is also of the form (3.26). Hence by (1) and (2) we get $Qw = 0$. We can see that the tau function satisfies an equation of Hirota type as follows. Making the change of variable $x_j = x_j + y_j$, $x_j' = x_j - y_j$, we get

$$\oint \frac{\mathrm{d}k}{2\pi \mathrm{i}} \exp\left(2\sum_{j=1}^{\infty} k^j y_j\right) \tau\left(x_1 + y_1 - \frac{1}{k}, x_2 + y_2 - \frac{1}{2k^2}, \ldots\right)$$

$$\times \tau\left(x_1 - y_1 + \frac{1}{k}, x_2 - y_2 + \frac{1}{2k^2}, \ldots\right)$$

$$= \oint \frac{\mathrm{d}k}{2\pi \mathrm{i}} \exp\left(2\sum_{j=1}^{\infty} k^j y_j\right) \exp\left(\sum_{j=1}^{\infty} \left(y_l - \frac{1}{lk^l}\right) \mathrm{D}_l\right) \tau \cdot \tau.$$

Thus expanding

$$\exp\left(2\sum_{j=1}^{\infty} k^j y_j\right) \exp\left(\sum_{j=1}^{\infty} \left(y_l - \frac{1}{lk^l}\right) \mathrm{D}_l\right) \qquad (3.27)$$

in $(y_1, y_2, \ldots)$ and taking the coefficient of k^{-1} gives an equation of Hirota type (see Exercise 3.7).

Exercises to Chapter 3

3.1. Find some solutions of (3.2) that are polynomials in x and t.
3.2. Verify that the 3-soliton solution of the KdV equation satisfies (3.2).
3.3. Prove (3.11).
3.4. Compute the product $X(p_1, q_1)X(p_2, q_2)$ and the commutator bracket $[X(p_1, q_1), X(p_2, q_2)]$, following the method of calculation of Lemma 3.1. Does the commutator vanish?
3.5. Find the tau function of the KP hierarchy as a polynomial in the three variables (x_1, x_2, x_3). In other words, you have to find a polynomial that satisfies (3.20).
3.6. Prove the assertion (2) concerning (3.26).
3.7. Prove that (3.13) follows from (3.27).

4
The calculus of Fermions

As we become more familiar with solitons and their structural properties, the algebraic laws governing the symmetry behind the equations come gradually to the fore. The scene changes for a while to this algebraic world; in this chapter we explain Fermions and their calculus.

4.1 The Bosonic algebra of differentiation and multiplication

We have treated infinitesimal transformations of differential equations (or functions) quite generally in terms of evolution equations. To deal with cases such as the KdV or KP equations having infinitely many symmetries, and to treat this hierarchy all at one go, we have naturally been led to considering functions of infinitely many variables $\mathbf{x} = (x_1, x_2, \ldots)$. For the sake of definiteness, for the moment, we restrict the functions we consider to polynomials in these variables. Although the number of variables is infinite, each polynomial itself is a finite sum of monomials, so involves only finitely many of the variables. When calculating the weight of a polynomial, we give each variable x_n the weight n.

Now when we speak of transformation of functions, in the first instance, we have the most basic operations of differentiation and multiplication. We define operators a_n and a_n^* acting on polynomials $f(\mathbf{x})$ by the following rules:

$$(a_n f)(\mathbf{x}) = \frac{\partial f}{\partial x_n}(\mathbf{x}), \quad (a_n^* f)(\mathbf{x}) = x_n f(\mathbf{x}). \tag{4.1}$$

We see at once that these operators satisfy the following commutation relations, called the *canonical commutation relations*

$$[a_m, a_n] = 0, \quad [a_m^*, a_n^*] = 0 \quad \text{and} \quad [a_m, a_n^*] = \delta_{mn}. \tag{4.2}$$

4.1 The Bosonic algebra of differentiation and multiplication

More generally, we can define natural operations of product and sum on differential operators with polynomial coefficients

$$\sum c_{\alpha_1\alpha_2\cdots\beta_1\beta_2\cdots} x_1^{\alpha_1} x_2^{\alpha_2} \cdots \left(\frac{\partial}{\partial x_1}\right)^{\beta_1} \left(\frac{\partial}{\partial x_2}\right)^{\beta_2} \cdots, \qquad (4.3)$$

which make the set of all of these into an algebra.

We want to consider $\{a_n, a_n^*\}_{n=1,2,\ldots}$ as a set of abstract symbols satisfying the relations (4.2); the a_n and a_n^* are called *Bosons*. Quite generally, given a set of letters S and a set of relations R holding between them, by starting from S and using the operations of product, sum and scalar multiplication an arbitrary finite number of times, we obtain *the algebra generated* by S with *defining relations* R. The algebra $\mathcal{B}$ generated by $S = \{a_n, a_n^*\}_{n=1,2,\ldots}$ with the defining relations (4.2) is called the *Heisenberg algebra*. We can see that by successively using the canonical commutation relations, we can express any element of $\mathcal{B}$ in a unique way as a linear combination of the following elements:

$$a_{m_1}^{*\alpha_1} \cdots a_{m_r}^{*\alpha_r} a_{n_1}^{\beta_1} \cdots a_{m_s}^{\beta_s}$$

for $m_1 < \cdots < m_r$, $n_1 < \cdots < n_s$ and $\alpha_i, \beta_j = 1, 2, \ldots$.

Quite generally, a *representation* of an algebra A on a vector space V is a linear map $\rho\colon A \to \mathrm{End}(V)$ satisfying $\rho(ab) = \rho(a)\rho(b)$ for all $a, b \in A$. If A has generators S and defining relations R, this is equivalent to specifying a linear map $\rho(s)$ of V for each $s \in S$ such that the relations R are satisfied.

According to this, formula (4.1) means that writing $\rho(a_n) = \partial/\partial x_n$ and $\rho(a_n^*) = x_n$ (multiplication by the variable x_n) defines a representation of the Heisenberg algebra $\mathcal{B}$ on the space of all polynomials $\mathbb{C}[\mathbf{x}] = \mathbb{C}[x_1, x_2, x_3, \ldots]$. The representation space $\mathbb{C}[\mathbf{x}]$ is called the *Bosonic Fock space*. The operators of differentiating $a_n = \partial/\partial x_n$ are called *annihilation operators* and those of multiplication $a_n^* = x_n$ *creation operators*. Notice that all the creation operators commute among themselves, as do all the annihilation operators.

The element $1 \in \mathbb{C}[\mathbf{x}]$ is called the *vacuum state*; then the following clearly hold.

(1) The annihilation operators φ kill the vacuum state: $\varphi 1 = 0$.
(2) The Fock space is generated by the vacuum state:

$$\mathbb{C}[\mathbf{x}] = \mathcal{B} \cdot 1 =_{\mathrm{def}} \{a \cdot 1 \mid a \in \mathcal{B}\}. \qquad (4.4)$$

In more detail, $\mathbb{C}[\mathbf{x}]$ has a basis

$$\{a^*_{m_1} \cdots a^*_{m_r} 1 \mid 0 < m_1 \leq \cdots \leq m_r\} \tag{4.5}$$

made up by allowing the creation operators to act on the vacuum state.

Maybe you object that we are deliberately saying simple things in an unnecessarily complicated way? In what follows, rather than as the concrete operation of differentiating and multiplication, Bosons will really appear in exactly the abstract form (4.2). Please bear with us for a moment.

4.2 Fermions

Along with the algebra $\mathcal{B}$, from now on we are mainly going to work with a different algebra $\mathcal{A}$, obtained from the above canonical commutation relations by replacing the commutator $[X, Y] = XY - YX$ by the *anticommutator*

$$[X, Y]_+ =_{\text{def}} XY + YX.$$

Definition 4.1 *We set up some symbols ψ_n, ψ_n^*, and the following basic relations holding between them, called the* canonical anticommutation relations. *We could take the indices n to be anything we like, but for our subsequent purposes we let n run through the half-integers $\mathbb{Z} + 1/2$.*

$$[\psi_m, \psi_n]_+ = 0, \quad [\psi_m^*, \psi_n^*]_+ = 0 \quad \text{and} \quad [\psi_m^*, \psi_n]_+ = \delta_{m+n, 0}. \tag{4.6}$$

Then ψ_n, ψ_n^* are called Fermions; *the algebra $\mathcal{A}$ they generate, with the defining relations (4.6), is called the* Clifford algebra. *Note that the relations (4.6) include as a particular case the characteristic property of Fermions*

$$\psi_n^2 = 0, \quad \psi_n^{*2} = 0.$$

As with the Bosons, by using (4.6) successively, we can transpose the order of products of expressions in the ψ_m, ψ_n^*, so that a general element of $\mathcal{A}$ can be written as a finite linear combination of monomials of the form

$$\psi_{m_1} \cdots \psi_{m_r} \psi_{n_1}^* \cdots \psi_{n_s}^*, \quad \text{where } m_1 < \cdots m_r \text{ and } n_1 < \cdots < n_s. \tag{4.7}$$

Remark 4.1 *Whether the algebra defined in this way is well defined and without contradiction is a question that merits closer scrutiny. The*

Table 4.1. *Bosons versus Fermions*

	Heisenberg algebra $\mathcal{B}$	Clifford algebra $\mathcal{A}$
generators	Bosons a, a^*	Fermions ψ, ψ^*
relations	$aa^* - a^*a = 1$	$\psi\psi^* + \psi^*\psi = 1, \psi^2 = \psi^{*2} = 0$
basis	$a^m a^{*n}$ for $m, n = 0, 1, 2, \ldots$	$1, \psi, \psi^*, \psi\psi^*$

precise argument is not very difficult, but we do not go into it, because it would take us away from main topic of this book; see for example Bourbaki, *Algèbre*, Chap. 9, §9. We only state the conclusion: the elements (4.7) are linearly independent and form a vector space basis of $\mathcal{A}$.

We summarise the similarities and differences between the Bosons and Fermions. For simplicity, we consider the case of just two generators a, a^* or ψ, ψ^*. What we get is tabulated in Table 4.1. More generally, the Clifford algebra $\mathcal{A}$ generated by any finite number of Fermions is finite dimensional, but the Heisenberg algebra $\mathcal{B}$ is infinite dimensional already in the two generator case.

Fermions can be realised in a concrete way using matrices. In the present case we need only set

$$\psi = \begin{pmatrix} 0 & 1 \\ 0 & 0 \end{pmatrix}, \quad \psi^* = \begin{pmatrix} 0 & 0 \\ 1 & 0 \end{pmatrix}$$

(compare Exercise 4.1). In contrast, Bosons can certainly not be written as matrices of any finite size: for if A, A^* were $n \times n$ matrices satisfying $AA^* - A^*A = 1$ then taking trace of both sides would give $0 = \text{Tr}(AA^*) - \text{Tr}(A^*A) = \text{Tr}(1) = n$, a contradiction.

4.3 The Fock representation

We now explain the Fermionic analogue of the Fock representation of Bosons on $\mathbb{C}[x_1, x_2, x_3, \ldots]$.

We consider diagrams made up of black and white Go stones lined up along the real line, indexed by half-integers; we require that far away to the right (when $n \gg 0$) all the stones are black, whereas far away to the left (when $n \ll 0$), they are all white. A diagram of this form is called a *Maya diagram* (see Figure 4.1).

By writing $m_1, m_2, \ldots$ for the positions of the black stones, we can

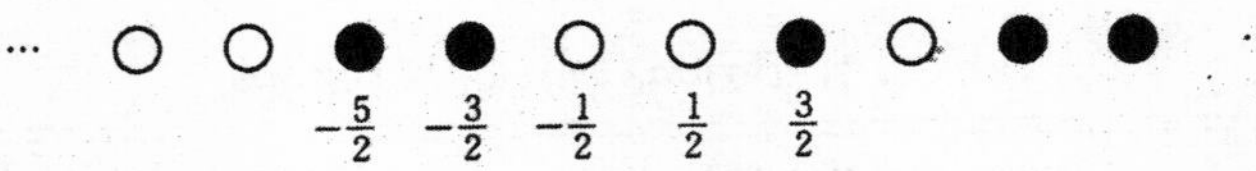

Fig. 4.1. A Maya diagram

describe a Maya diagram as an increasing sequence of half-integers

$$\mathbf{m} = \{m_j\}_{j \geq 1} \quad \text{with } m_1 < m_2 < m_3 < \cdots, \tag{4.8}$$

and $m_{j+1} = m_j + 1$ for all sufficiently large j. We set $\mathcal{F}$ to be the vector space based by the set of Maya diagrams, and call it the *Fermionic Fock space*. The basis vector corresponding to the Maya diagram (4.8) is written $|\mathbf{m}\rangle$. We determine a left action of the Fermions on Fock space by the following rules:

$$\psi_n|\mathbf{m}\rangle = \begin{matrix} (-1)^{i-1}|\ldots, m_{i-1}, m_{i+1}, \ldots\rangle & \text{if } m_i = -n \text{ for some } i, \\ 0 & \text{otherwise;} \end{matrix} \tag{4.9}$$

$$\psi_n^*|\mathbf{m}\rangle = \begin{matrix} (-1)^i|\ldots, m_i, n, m_{i+1}, \ldots\rangle & \text{if } m_i < n < m_i + 1 \text{ for some } i, \\ 0 & \text{otherwise;} \end{matrix}$$
(4.10)

except that in the case $i = 1$, we must obviously interpret (4.9) as giving $|m_2, m_3, \ldots\rangle$ and when $i = 0$, we interpret (4.10) as giving $|n, m_1, m_2, \ldots\rangle$. You should check that under this definition, the canonical anticommutation relations (4.6) are satisfied. The Fermion ψ_n performs the role of creating a white stone at $-n$ (or equivalently, annihilating a black stone there), whereas ψ_n^* creates a black stone at n (or equivalently, annihilates a white stone there).

We divide up the Fermions into two classes:

$\{\psi_n, \psi_n^*\}$ for $n < 0$ are called *creation operators*;
$\{\psi_n, \psi_n^*\}$ for $n > 0$ are called *annihilation operators*.

Then all the creation operators anticommute among themselves, as do all the annihilation operators. (We say that X and Y anticommute if $XY = -YX$.) Now consider the vector corresponding to the diagram with the entire left half-line $n < 0$ filled with white stones, and the entire right half-line $n > 0$ with black stones, so that $m_j = j - 1/2$ for $j = 1, 2, \ldots$; we write $|\text{vac}\rangle$ for this vector, and call it the *vacuum state*. Then the following properties hold.

(1) The annihilation operators φ kill the vacuum state: $\varphi|\text{vac}\rangle = 0$.

4.4 Duality, charge and energy

(2) The Fock space is generated by the vacuum state:

$$\mathcal{F} = \mathcal{A} \cdot |\text{vac}\rangle =_{\text{def}} \{a|\text{vac}\rangle \mid a \in \mathcal{A}\}.$$

In fact it is known that the Fock space is characterised by these two properties. By successively applying creation operators to the vacuum state we obtain (up to sign) the vector corresponding to any Maya diagram:

$$\psi_{m_1} \cdots \psi_{m_r} \psi^*_{n_1} \cdots \psi^*_{n_s} |\text{vac}\rangle \qquad (4.11)$$
$$\text{for } m_1 < \cdots < m_r < 0 \text{ and } n_1 < \cdots < n_s < 0.$$

The elements (4.11) are linearly independent and provide a basis of $\mathcal{F}$.

Example 4.1 *The following discussion may help to understand the sign convention in the Fock representation. Suppose we could set up the 'fake vacuum state' $|\Omega\rangle$, which is killed by all the ψ_n for $n \in \mathbb{Z}$. Then the 'genuine' vacuum state $|\text{vac}\rangle$ can be thought of formally as the vector obtained by successively applying the infinitely many operators ψ^*_n to $|\Omega\rangle$:*

$$|\text{vac}\rangle = \psi^*_{1/2} \psi^*_{3/2} \psi^*_{5/2} \cdots |\Omega\rangle.$$

In fact the expression on the right-hand side is killed exactly by all the annihilation operators ψ_n, ψ^*_n for $n > 0$.

The corresponding picture of the vacuum in elementary particle physics is called Dirac's sea: *the vacuum state is not the state having no particles at all, but that in which the antiparticles are in a certain definite state, filling up all the 'holes'.*

If we apply further creation operators to this, we get

$$\begin{aligned}
\psi_{-3/2}|\text{vac}\rangle &= \psi_{-3/2}\psi^*_{1/2}\psi^*_{3/2}\psi^*_{5/2}\cdots|\Omega\rangle \\
&= -\psi^*_{1/2}\psi_{-3/2}\psi^*_{3/2}\psi^*_{5/2}\cdots|\Omega\rangle \\
&= -\psi^*_{1/2}\psi^*_{5/2}\psi^*_{7/2}\cdots|\Omega\rangle, \quad \text{and} \\
\psi^*_{-3/2}|\text{vac}\rangle &= \psi^*_{-3/2}\psi^*_{1/2}\psi^*_{3/2}\psi^*_{5/2}\cdots|\Omega\rangle,
\end{aligned}$$

and so on. This can be pictured as in Figure 4.2 as the annihilation of a black stone (equivalently, the creation of a white stone), or its inverse.

4.4 Duality, charge and energy

We define the dual Fock space $\mathcal{F}^*$ in parallel with $\mathcal{F}$. In this case, we describe the Maya diagram using the positions $\ldots, n_3, n_2, n_1$ of the white

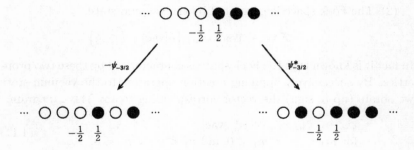

Fig. 4.2. Creation and annihilation operators

stones. (Here $n_i \in \mathbb{Z} + 1/2$.) As a vector space $\mathcal{F}^*$ is again based by the Maya diagrams, with the basis vectors written as

$$\langle \mathbf{n} | = \langle \ldots, n_3, n_2, n_1 |, \quad \text{where } \ldots < n_3 < n_2 < n_1$$

and $n_{j+1} = n_j - 1$ for all sufficiently large j. We define a right action of the Fermions on $\mathcal{F}^*$ as follows:

$$\langle \mathbf{n} | \psi_n = \begin{cases} (-1)^i \langle \ldots, n_{i+1}, n, n_i, \ldots | & \text{if } n_{i+1} < n < n_i \text{ for some } i, \\ 0 & \text{otherwise}; \end{cases} \quad (4.12)$$

$$\langle \mathbf{n} | \psi_n^* = \begin{cases} (-1)^{i-1} \langle \ldots, n_{i+1}, n_{i-1}, \ldots | & \text{if } n = -n_i \text{ for some } i, \\ 0 & \text{otherwise}. \end{cases} \quad (4.13)$$

The dual vacuum state $\langle \text{vac} |$ is the $\langle \mathbf{n} |$ defined by $n_j = -j + 1/2$ for $j = 1, 2, \ldots$, and just as for the Fock space, we have

(1) the annihilation operators φ kill the vacuum state $-\langle \text{vac} | \varphi = 0$;
(2) the Fock space is generated by the vacuum state:

$$\mathcal{F}^* = \langle \text{vac} | \cdot \mathcal{A} =_{\text{def}} \{ a | \text{vac} \rangle \mid a \in \mathcal{A} \}.$$

We can take

$$\langle \text{vac} | \psi_{m_1} \cdots \psi_{m_r} \psi_{n_1}^* \cdots \psi_{n_s}^*$$
for $0 < m_1 < \cdots < m_r$ and $0 < n_1 < \cdots < n_2$

as a basis of $\mathcal{F}^*$. We have been using Dirac's bra and ket notation $\langle u |$ for an element of $\mathcal{F}^*$ and $|u\rangle$ for an element of $\mathcal{F}$, which is traditional in quantum physics; the action of $a \in \mathcal{A}$ on elements of $\mathcal{F}^*$ and $\mathcal{F}$ is written $\langle u | a$ or $a | u \rangle$. There is a pairing $\mathcal{F}^* \times \mathcal{F} \to \mathbb{C}$ denoted by $(\langle u |, |v\rangle) \mapsto \langle u | v \rangle$ between the two spaces, defined by the formula

$$\langle \mathbf{n} | \mathbf{m} \rangle = \delta_{m_1+n_1, 0} \delta_{m_2+n_2, 0} \delta_{m_3+n_3, 0} \cdots, \quad (4.14)$$

4.4 Duality, charge and energy

Table 4.2. *Charge and energy of a Fermion*

Fermion	ψ_n	ψ_n^*
charge	1	-1
energy	$-n$	$-n$

where $\langle \mathbf{n} | = \langle \cdots n_3 n_2 n_1 |$ and $|\mathbf{m}\rangle = |m_1 m_2 m_3 \cdots \rangle$. Then the following properties hold:

$$\langle \text{vac} | \text{vac} \rangle = 1 \quad \text{and} \quad ((\langle u | a) | v \rangle = \langle u | (a | v \rangle) \quad \text{for all } a \in \mathcal{A}. \quad (4.15)$$

We write $\langle u | a | v \rangle$ for the latter expression.

The (electric) charge and energy of the Fermions ψ_n and ψ_n^* are defined in Table 4.2. More generally, the charge and energy of a monomial in the ψ and ψ^* is the sum of those of the factors, so that by definition, the monomial (4.7) has charge $r-s$ and energy $-(m_1 + \cdots + m_r + n_1 + \cdots + n_s)$.

Charge and energy are also defined for the basis elements $|u\rangle$ of the Fock space $\mathcal{F}$. For this, we set

charge (or energy) of $|\text{vac}\rangle = 0$,

charge (or energy) of $a|\text{vac}\rangle = $ that of a.

Here a is a monomial in the Fermions for which $a|\text{vac}\rangle \neq 0$. Similarly, for the dual Fock space $\mathcal{F}^*$, we set

charge (or energy) of $\langle\text{vac}| = 0$,

charge (or energy) of $\langle\text{vac}|a = -$that of a

indexcharge(note the minus sign). We write $\mathcal{F}_l^{(d)}$ for the vector subspace of the Fock space generated by basis vectors with definite charge l and energy d, that is,

$\mathcal{F}_l^{(d)} = $ linear span of

$$\left\{ \psi_{m_1} \cdots \psi_{m_r} \psi_{n_1}^* \cdots \psi_{n_s}^* |\text{vac}\rangle \,\middle|\, \begin{array}{l} m_1 < \cdots < m_r, n_1 < \cdots n_s < 0, \\ r - s = l \text{ and } \sum m_i + \sum n_j = d \end{array} \right\}.$$

Then $\mathcal{F}$ decomposes as a direct sum of vector spaces $\mathcal{F} = \bigoplus_l \mathcal{F}_l$, and moreover, $\mathcal{F}_l = \bigoplus_d \mathcal{F}_l^{(d)}$. The same thing holds for the dual Fock space.

Now for every integer l, consider the Maya diagram obtained by sliding the diagram for the vacuum state bodily l steps to the right (that is, $-l$ steps to the left if $l < 0$); we write $|l\rangle = |l + 1/2, l + 3/2, l + 5/2, \ldots\rangle$

for the corresponding vector of $\mathcal{F}$. In the same way (swapping left and right), we define the element $\langle l| = \langle \ldots, l-5/2, l-3/2, l-1/2| \in \mathcal{F}^*$. In other words, $|l\rangle$ and $\langle l|$ are defined as follows:

$$\langle l| = \begin{cases} \langle \mathrm{vac}|\psi_{1/2}\cdots\psi_{-l-1/2} & \text{for } l < 0, \\ \langle \mathrm{vac}| & \text{for } l = 0, \\ \langle \mathrm{vac}|\psi^*_{1/2}\cdots\psi^*_{l-1/2} & \text{for } l > 0; \end{cases}$$

$$|l\rangle = \begin{cases} \psi^*_{l+1/2}\cdots\psi^*_{-1/2}|\mathrm{vac}\rangle & \text{for } l < 0, \\ |\mathrm{vac}\rangle & \text{for } l = 0, \\ \psi_{-l+1/2}\cdots\psi_{-1/2}|\mathrm{vac}\rangle & \text{for } l > 0. \end{cases}$$

It is clear that among vectors of definite charge l, these are the vectors having minimum energy $d = l^2/2$. Note that by definition, we have

$$\langle l|\psi_n = 0 \quad \text{for } n < -l \quad \text{and} \quad \langle l|\psi^*_n = 0 \quad \text{for } n < l; \qquad (4.16)$$
$$\psi_n|l\rangle = 0 \quad \text{for } n > -l \quad \text{and} \quad \psi^*_n|l\rangle = 0 \quad \text{for } n > l. \qquad (4.17)$$

4.5 Wick's theorem

We now explain how the pairing (4.14) between Fock space and its dual can be determined uniquely from its properties (4.15), without any reference to the concrete definition. In what follows, we define the *vacuum expectation value* of $a \in \mathcal{A}$ to be the number $\langle \mathrm{vac}|a|\mathrm{vac}\rangle$, which we abbreviate to $\langle a \rangle$. Then (4.15) and the definition of creation and annihilation operators imply at once the following properties:

$$\left.\begin{array}{l} \langle 1 \rangle = 1, \quad \langle \psi_n \rangle = 0, \quad \langle \psi^*_n \rangle = 0, \\ \langle \psi_m \psi_n \rangle = 0, \quad \langle \psi^*_m \psi^*_n \rangle = 0, \quad \langle \psi_m \psi^*_n \rangle = \delta_{m+n,0}\theta(n < 0). \end{array}\right\} \qquad (4.18)$$

Here the notation $\theta(P)$ in the final term is the Boolean characteristic function of a general property P, equal to 1 if P is true, 0 otherwise.

For example, the final equation in (4.18) works out as follows: the right-hand side can obviously only be nonzero if $n < 0 < m$. But then by (4.6), the left-hand side equals

$$\langle \delta_{m+n,0} - \psi_n \psi^*_m \rangle = \delta_{m+n,0}\langle 1 \rangle.$$

Continuing the calculation in a similar vein gives the following:

$$\begin{aligned} \langle \psi_k \psi_m \psi^*_n \rangle &= 0, \\ \langle \psi_k \psi_l \psi^*_m \psi^*_n \rangle &= \langle \psi_k \psi^*_n \rangle \langle \psi_l \psi^*_m \rangle - \langle \psi_k \psi^*_m \rangle \langle \psi_l \psi^*_n \rangle, \\ &\cdots. \end{aligned}$$

The general pattern is easy to figure out from this method of calculating: using the commutation relation (4.6), we can successively pass all the annihilation operators over to the right and the creation operators to the left; either of these gives 0 when it bumps into the vacuum vector at the ends. However, each time we swap two Fermions over, the constant term in the commutator generates a little small change; adding up all the resulting constant terms gives the vacuum expectation value.

Thus we see that any monomial a in the ψ and ψ^* has $\langle a \rangle = 0$ unless the ψ and ψ^* occur the same number of times. Therefore $\mathcal{F}_k^*$ and $\mathcal{F}_l$ are orthogonal if $k \neq l$. More precisely, $\mathcal{F}_k^{*(d)}$ and $\mathcal{F}_l^{(e)}$ are orthogonal unless $k = l$ and $d = e$.

When the number of occurrences is equal, we have to take account of all the combinatorial possibilities for the positions of ψ^* and ψ. Just a little care is needed with the sign change that occurs each time we swap two Fermions over. We summarise as a theorem the result of calculation just described. We write $W = (\bigoplus_{n \in \mathbb{Z}} \mathbb{C}\psi_n) \oplus (\bigoplus_{n \in \mathbb{Z}} \mathbb{C}\psi_n^*)$ for the set of all linear combinations of Fermions.

Theorem 4.1 (Wick's theorem) *For $w_1, \ldots, w_r \in W$ we have*

$$\langle w_1 \cdots w_r \rangle = \begin{cases} 0 & \text{if } r \text{ is odd,} \\ \sum_\sigma \text{sign}(\sigma) \langle w_{\sigma(1)} w_{\sigma(2)} \rangle \cdots \langle w_{\sigma(r-1)} w_{\sigma(r)} \rangle & \text{if } r \text{ is even,} \end{cases}$$

where $\text{sign}(\sigma)$ is the sign of a permutation; the sum runs over all permutations σ satisfying $\sigma(1) < \sigma(2), \ldots, \sigma(r-1) < \sigma(r)$ and $\sigma(1) < \sigma(3) < \cdots < \sigma(r-1)$, in other words, over all ways of grouping the w_i into pairs.

Exercises to Chapter 4

4.1. Consider the Clifford algebra generated by two elements $\{\psi, \psi^*\}$ related by $\psi^2 = \psi^{*2} = 0$ and $[\psi, \psi^*]_+ = 0$. Define the Fock representation $\mathcal{F}$ as in the text, with a vacuum state $|\text{vac}\rangle \in \mathcal{F}$ for which $\psi|\text{vac}\rangle = 0$. Prove that $v_1 = |\text{vac}\rangle$, $v_2 = \psi^*|\text{vac}\rangle$ forms a basis of $\mathcal{F}$ in which the action of ψ and ψ^* is given by the matrices

$$\psi \leftrightarrow \begin{pmatrix} 0 & 1 \\ 0 & 0 \end{pmatrix}, \quad \psi^* \leftrightarrow \begin{pmatrix} 0 & 0 \\ 1 & 0 \end{pmatrix}.$$

4.2. Write out Wick's theorem correctly in the case $n = 6$.

4.3. Prove that $\langle \psi_{m_1} \cdots \psi_{m_s} \psi_{n_s}^* \cdots \psi_{n_1}^* \rangle = \det\left(\langle \psi_{m_i} \psi_{n_j}^* \rangle\right)$. Using this, prove that (4.15) is a nondegenerate pairing on $\mathcal{F}_l^* \times \mathcal{F}_l$. Here we say that a bilinear form $F: V \times W \to \mathbb{C}$ is *nondegenerate* if

$$F(v,w) = 0 \quad \text{for all } w \in W \Longrightarrow v = 0,$$
$$F(v,w) = 0 \quad \text{for all } v \in V \Longrightarrow w = 0.$$

5
The Boson–Fermion correspondence

Although the construction of Bosons and Fermions in the preceding chapter proceeded along parallel lines, in character the two are remarkably different. Despite this, the main theme of this chapter is that we can actually realise each of them in terms of the other. To pull off this kind of stunt, the essential idea is to make use of infinite sums of Bosons and Fermions. The generating functions we introduce provide a glimpse of the atmosphere of quantum field theory.

5.1 Using generating functions

We now explain the idea of *generating functions*, an important tool in gaining insight into how to proceed systematically with calculations. Introducing a variable k, we define the *Fermionic generating functions* as the formal sums

$$\psi(k) = \sum_{n\in\mathbb{Z}+1/2} \psi_n k^{-n-1/2} \quad \text{and} \quad \psi^*(k) = \sum_{n\in\mathbb{Z}+1/2} \psi_n^* k^{-n-1/2}. \quad (5.1)$$

In what follows, we frequently work with identities between formal sums; for example, an identity of the form $\sum a_n k^n = \sum b_n k^n$ is interpreted as a means of writing out and grouping together the series of identities $a_n = b_n$ between the coefficients.

Example 5.1 *Let us calculate the vacuum expectation value of the generating function (5.1). Using (4.18), we get*

$$\langle \psi(p)\psi^*(q)\rangle = \sum_{m\in\mathbb{Z}+1/2}\sum_{n\in\mathbb{Z}+1/2} \langle \psi_m \psi_n^*\rangle p^{-m-1/2} q^{-n-1/2}$$

$$= \sum_{n=0}^{\infty} p^{-n-1} q^n. \quad (5.2)$$

The final expression can be written

$$\langle \psi(p)\psi^*(q) \rangle = \frac{1}{p-q}. \qquad (5.3)$$

However impressive the formula may look, it means neither more nor less than (5.2); in other words, the right-hand side of (5.3) is interpreted as an expansion in p,q with $|p| > |q|$.

Example 5.2 *More generally, by Wick's theorem, we have*

$$\begin{aligned}\langle \psi(p_1)\cdots\psi(p_n)\psi^*(q_n)\cdots\psi^*(q_1)\rangle &= \det\left(\langle \psi(p_i)\psi^*(q_j)\rangle\right) \\ &= \det\left(\frac{1}{p_i - q_j}\right).\end{aligned}$$

Factorising the right-hand side as a rational function gives the formula

$$\langle \psi(p_1)\cdots\psi(p_n)\psi^*(q_n)\cdots\psi^*(q_1)\rangle = \frac{\prod_{1\leq i<j\leq n}(p_i - p_j)(q_j - q_i)}{\prod_{1\leq i<j\leq n}(p_i - q_j)}. \qquad (5.4)$$

Here again, (5.4) is understood as an expansion in p_i, q_j with $|p_1| > \cdots > |p_n| > |q_n| > \cdots > |q_1|$.

Remark 5.1 *The factorisation (5.4) of the determinant $\det\left(\frac{1}{p_i - q_j}\right)$ is called* Cauchy's identity. *To prove it, note that if we multiply the determinant by $\prod_{1\leq i<j\leq n}(p_i - q_j)$, the resulting polynomial has zeros along all the diagonals $p_i = p_j$ and $q_i = q_j$ for $i \neq j$, and so is divisible by $\prod_{1\leq i<j\leq n}(p_i - p_j)(q_j - q_i)$. After this, we need only compare degrees.*

5.2 The normal product

When handling differential operators, we usually write them out (perhaps even unconsciously) in a 'normal' order, with differentiation on the right, and multiplication by functions on the left. One reason for doing this is that the expression is not well defined until we choose some order (for example, $(\partial/\partial x_n)x_n = x_n(\partial/\partial x_n) + 1$; moreover, specifying the order as above makes operators meaningful even if they include infinite sums.

Example 5.3 *Consider the Euler operator $\sum_{n=1}^{\infty}(nx_n\partial/\partial x_n)$, which takes a weighted homogeneous polynomial $f(\mathbf{x})$ into $d \times f(\mathbf{x})$, where $d = \deg_{\mathbf{x}} f$ (recall that the x_n are weighted with $\deg x_n = n$); it is*

5.2 The normal product

thus the operator that measures weights. It is meaningful applied to any polynomial, although it involves an infinite sum. If instead we swap the order and try writing $\sum_{n=1}^{\infty} n(\partial/\partial x_n) x_n$, the new operator is meaningless, since already applied to the constant function 1, it would give $\sum_{n=1}^{\infty} n(1 + x_n \partial/\partial x_n) \cdot 1 = \infty$.

We now introduce a notation which will be convenient. Quite generally, for a polynomial p in the operations of differentiation and multiplication, the colon notation $:p:$ is defined inductively as follows:

(1) $:p:$ is a linear function of p, and all the differentiation and multiplication operators within the colons commute among themselves;

(2) $:1: = 1$, $:p \cdot \dfrac{\partial}{\partial x_n}: = :p: \dfrac{\partial}{\partial x_n}$ and $:x_n \cdot p: = x_n :p:$.

The colon notation $:p:$ is called the normally ordered product of p, or simply the *normal product* of p.

Example 5.4 We have $:x_n \dfrac{\partial}{\partial x_n}: = :\dfrac{\partial}{\partial x_n} x_n: = x_n \dfrac{\partial}{\partial x_n}$; and

$$:e^{x_n + \partial/\partial x_n}: = e^{x_n} e^{\partial/\partial x_n}.$$

Strictly speaking, letting e^{x_n} act on polynomials takes us beyond polynomials; however, $e^{\partial/\partial x_n}$ acting on polynomials has a genuine meaning, and it is easy to see that it is the translation operator $f(\mathbf{x}) \mapsto f(\ldots, x_n + 1, \ldots)$.

In parallel with this, we now introduce the Fermionic normal product for elements of the Clifford algebra $a \in \mathcal{A}$. We use the same colon notation $:\ :$ as for the Bosonic normal product; however, in cases of ambiguity, we distinguish the two by a subscript, $:\ :_B$ (for Bosonic) and $:\ :_F$ (for Fermionic). The axioms are as follows:

(1) $:a:$ is a linear function of a, and all the Fermions within the colons anticommute among themselves;

(2) $:1: = 1$, and

$$\begin{cases} :a \cdot \varphi: = :a: \varphi & \text{for } \varphi \text{ an annihilation operator,} \\ :\varphi^* \cdot a: = \varphi^* :a: & \text{for } \varphi^* \text{ a creation operator.} \end{cases}$$

For example, for a quadratic monomial in Fermions we have

$$:\psi_m \psi_n^*: = \begin{cases} \psi_m \psi_n^* & \text{if } m < 0 \text{ or if } n > 0, \\ -\psi_n^* \psi_m & \text{if } m > 0 \text{ or if } n < 0, \end{cases}$$
$$= \psi_m \psi_n^* - \langle \psi_m \psi_n^* \rangle. \tag{5.5}$$

The important point for us is that expressions including infinite sums make sense as operators on Fock space, provided we write them as normal products. We will see an example of this in the next section.

5.3 Realising the Bosons

We define operators H_n, one for each integer $n \in \mathbb{Z}$, by introducing the generating function

$$\sum_{n \in \mathbb{Z}} H_n k^{-n-1} = \; : \psi(k)\psi^*(k) : . \tag{5.6}$$

Comparing coefficients of the k^n gives

$$H_n = \sum_{j \in \mathbb{Z}+1/2} : \psi_{-j}\psi^*_{j+n} : . \tag{5.7}$$

Now any element $|u\rangle$ of the Fock space is obtained by successively applying Fermions to the vacuum state $|\text{vac}\rangle$, and it follows that $: \psi_{-j}\psi^*_{j+n} : |u\rangle = 0$ for all but finitely many j. Thus for any given $|u\rangle$, the expression $H_n|u\rangle$ is actually a finite sum.

Let us determine the commutation relations among the operators (5.7). Note first the following identities among commutator and anticommutator brackets:

$$[AB, C] = A[B, C]_+ - [A, C]_+ B \tag{5.8}$$
$$= A[B, C] + [A, C]B. \tag{5.9}$$

Calculating using (5.8) gives the following commutation relations with the Fermions:

$$[H_n, \psi_m] = \psi_{m+n}, \quad [H_n, \psi^*_m] = -\psi^*_{m+n}. \tag{5.10}$$

Now applying (5.9) and (5.10), we see that

$$[H_m, H_n] = m\delta_{m+n,0} \tag{5.11}$$

(see Exercise 5.1). Now all the commutators are numbers, so that these look just like the Bosonic commutation relations. In fact if for $n = 1, 2, \ldots$ we set $a_n = H_n$ and $na^*_n = H_{-n}$ then (5.11) coincides with the canonical commutation relations (4.2). Thus by allowing infinite sums, we have realised the Bosons in terms of Fermions!

The operator H_0 is different; alone among the operators (5.7), it commutes with all the H_n. However, if we set $n = 0$ and apply (5.10) successively, we find that H_0 is the operator measuring charge, in the sense that

$$a \text{ has charge equal to } l \iff [H_0, a] = la.$$

5.4 Isomorphism of Fock spaces

By definition, H_n has charge 0 and energy $-n$. Note in particular that each H_n takes the subspace $\mathcal{F}_l \subset \mathcal{F}$ of definite charge l into itself. We now want to explain how each of the $\mathcal{F}_l$ can be identified in a natural way with the *Bosonic* Fock space $\mathbb{C}[x_1, \ldots, x_n]$. To handle all the charges at one go we introduce a new variable z, and consider the space

$$\mathbb{C}[z, z^{-1}, x_1, x_2, x_3, \ldots] = \oplus_{l \in \mathbb{Z}} z^l \mathbb{C}[x_1, x_2, x_3, \ldots]. \tag{5.12}$$

We also define

$$H(\mathbf{x}) = \sum_{n=1}^{\infty} x_n H_n. \tag{5.13}$$

Note that by the rule (4.17), for every l we have

$$H(\mathbf{x})|l\rangle = 0, \tag{5.14}$$

where $|l\rangle$ was discussed in Section 4.4.

Now corresponding to an element $|u\rangle \in \mathcal{F}$, we define the following polynomial of $z^{\pm 1}$ and $\mathbf{x}$:

$$\Phi(|u\rangle) = \sum_{l \in \mathbb{Z}} z^l \langle l | e^{H(\mathbf{X})} | u \rangle. \tag{5.15}$$

If $|u\rangle$ has definite charge m, only the term with $l = m$ survives on the right-hand side. The right-hand side actually defines a polynomial; we will see presently how this works in a concrete example, but we start with the statement.

Theorem 5.1 (Isomorphism of Fock spaces) *The correspondence*

$$\Phi: \mathcal{F} \to \mathbb{C}[z, z^{-1}, x_1, x_2, x_3, \ldots], \quad \text{given by} \quad |u\rangle \mapsto \Phi(|u\rangle), \tag{5.16}$$

is an isomorphism of vector spaces. Moreover, we have

$$\Phi(H_n|u_n\rangle) = \left. \begin{array}{ll} \frac{\partial}{\partial x_n} \Phi(|u\rangle) & \text{if } n > 0, \\ -n x_{-n} \Phi(|u\rangle) & \text{if } n < 0. \end{array} \right\} \tag{5.17}$$

Proof We observe that the H_n for $n > 0$ commute among themselves, so that

$$\frac{\partial}{\partial x_n}\langle l|e^{H(\mathbf{x})}|u\rangle = \langle l|\frac{\partial}{\partial x_n}e^{H(\mathbf{x})}|u\rangle$$
$$= \langle l|e^{H(\mathbf{x})}H_n|u\rangle,$$

which gives the first line of (5.17). Moreover, in view of

$$\langle l|e^{H(\mathbf{x})}H_{-n}|u\rangle = \langle l|e^{H(\mathbf{x})}H_{-n}e^{-H(\mathbf{x})}e^{H(\mathbf{x})}|u\rangle, \qquad (5.18)$$

using $[H(\mathbf{x}), H_{-n}] = nx_n$, we see that

$$e^{H(\mathbf{x})}H_{-n}e^{-H(\mathbf{x})} = H_{-n} + [H(\mathbf{x}), H_{-n}] + \frac{1}{2!}[H(\mathbf{x}), [H(\mathbf{x}), H_{-n}]] + \cdots$$
$$= H_{-n} + nx_n.$$

However, (4.16) gives $\langle l|H_{-n} = 0$. Therefore the right-hand side of (5.18) becomes

$$\langle l|(H_{-n} + nx_n)e^{H(\mathbf{x})}|u\rangle = nx_n\langle l|e^{H(\mathbf{x})}|u\rangle,$$

and the second line of (5.17) follows.

In particular for $|u\rangle = |l\rangle$, (5.14) gives $\Phi(|l\rangle) = z^l$. Successively applying H_{-n} to this, we can obtain z^l times any monomial in the form $\Phi(|u\rangle)$. This says that (5.17) is surjective. The fact that it is injective follows by a degree counting argument, which we omit (see Exercise 5.5). Q.E.D.

We want to try to see in a particular example which polynomials each individual vector of the Fock space gets mapped to under the above map (5.17). From the definition (5.13) and the commutation relation (5.10), we get

$$[H(\mathbf{x}), \psi(k)] = \xi(\mathbf{x}, k)\psi(k) \quad \text{and} \quad [H(\mathbf{x}), \psi^*(k)] = -\xi(\mathbf{x}, k)\psi^*(k),$$

where $\xi(\mathbf{x}, k) = \sum_{n=1}^{\infty} k^n x_n$ (compare (2.18)). Thus a computation similar to the one after (5.18) gives

$$\left.\begin{array}{l} e^{H(\mathbf{x})}\psi(k)e^{-H(\mathbf{x})} = e^{\xi(\mathbf{x},k)}\psi(k), \\ e^{H(\mathbf{x})}\psi^*(k)e^{-H(\mathbf{x})} = e^{-\xi(\mathbf{x},k)}\psi^*(k). \end{array}\right\} \qquad (5.19)$$

Now if we set

$$e^{\xi(\mathbf{x},k)} = \sum_{n=0}^{\infty} p_n(\mathbf{x})k^n,$$

$$= 1 + x_1 k + \left(x_2 + \frac{x_1^2}{2}\right)k^2 + \left(x_3 + x_2 x_1 + \frac{x_1^3}{6}\right)k^3 + \cdots, \quad (5.20)$$

then from (5.19) and (5.20), together with the equations obtained by interchanging $\mathbf{x} \mapsto -\mathbf{x}$ in these, we get

$$e^{H(\mathbf{x})}\psi_n e^{-H(\mathbf{x})} = \sum_{j=0}^{\infty} \psi_{n+j} p_j(\mathbf{x})$$

$$= \psi_n + x_1\psi_{n+1} + \left(x_2 + \frac{x_1^2}{2}\right)\psi_{n+2} + \cdots, \qquad (5.21)$$

$$e^{H(\mathbf{x})}\psi_n^* e^{-H(\mathbf{x})} = \sum_{j=0}^{\infty} \psi_{n+j}^* p_j(-\mathbf{x})$$

$$= \psi_n^* - x_1\psi_{n+1}^* + \left(-x_2 + \frac{x_1^2}{2}\right)\psi_{n+2}^* + \cdots, \qquad (5.22)$$

and we can determine the polynomial represented by $\Phi(|u\rangle)$ by using these.

Example 5.5 For the vector $\psi_{-5/2}|\text{vac}\rangle$, we have

$$\Phi(\psi_{-5/2}|\text{vac}\rangle) = z\langle 1|e^{H(\mathbf{x})}\psi_{-5/2}|\text{vac}\rangle$$
$$= z\langle\text{vac}|\psi_{1/2}^* e^{H(\mathbf{x})}\psi_{-5/2}e^{-H(\mathbf{x})}|\text{vac}\rangle$$
$$= z \times \left(x_2 + \frac{x_1^2}{2}\right),$$

where we used (5.21) for the final equality.

In the same way, for the vector $\psi_{-3/2}\psi_{-3/2}^*|\text{vac}\rangle$, we get

$$\Phi(\psi_{-3/2}\psi_{-3/2}^*|\text{vac}\rangle) = \langle\text{vac}|e^{H(\mathbf{x})}\psi_{-3/2}e^{-H(\mathbf{x})}e^{H(\mathbf{x})}\psi_{-3/2}^* e^{-H(\mathbf{x})}|\text{vac}\rangle$$
$$= \langle\text{vac}| \left(\psi_{-3/2} + x_1\psi_{-1/2} + \left(x_2 + \frac{x_1^2}{2}\right)\psi_{1/2} + \cdots\right)$$
$$\times \left(\psi_{-3/2}^* - x_1\psi_{-1/2}^* + \cdots\right)|\text{vac}\rangle$$
$$= \left(x_3 + x_2 x_1 + \frac{x_1^3}{6}\right)\cdot 1 + \left(x_2 + \frac{x_1^2}{2}\right)\cdot(-x_1)$$
$$= x_3 - \frac{x_1^3}{3}.$$

The procedure involved in carrying out the calculation makes it clear that the answer is quite generally a polynomial.

5.5 Realising the Fermions

Theorem 5.1 shows that we can identify the Fermionic Fock space with the Bosonic Fock space. Thus it should be possible to realise the action of the Fermions on the former in terms of operators on the latter. We now set about achieving this.

Slightly out of the blue, we introduce the operators k^{H_0} and e^K on the space (5.12) by the formulas

$$(k^{H_0}f)(z,\mathbf{x}) =_{\mathrm{def}} f(kz,\mathbf{x}) \quad \text{and} \quad (\mathrm{e}^K f)(z,\mathbf{x}) =_{\mathrm{def}} zf(z,\mathbf{x}),$$

then define

$$\left.\begin{array}{l}\Psi(k) = \mathrm{e}^{\xi(\mathbf{X},k)}\mathrm{e}^{-\xi(\widetilde{\partial},k^{-1})}\mathrm{e}^K k^{H_0}, \\ \Psi^*(k) = \mathrm{e}^{-\xi(\mathbf{X},k)}\mathrm{e}^{\xi(\widetilde{\partial},k^{-1})}\mathrm{e}^{-K} k^{-H_0},\end{array}\right\} \quad (5.23)$$

where we have set

$$\widetilde{\partial} = \left(\frac{\partial}{\partial x_1}, \frac{1}{2}\frac{\partial}{\partial x_2}, \frac{1}{3}\frac{\partial}{\partial x_3}\right) \quad \text{and} \quad \xi(\widetilde{\partial},k^{-1}) = \sum_{n=1}^{\infty}\frac{1}{n}\frac{\partial}{\partial x_n}k^{-n}.$$

As we said above, H_0 is the operator measuring charge, so that $k^{H_0}z^l = k^l z^l$, and k^{H_0} is a natural notation to use. Writing e^K is more unexpected, but it can be motivated as follows. We formally define $\varphi(k)$ to be the series

$$\varphi(k) = \sum_{n\neq 0}\frac{H_n}{-n}k^{-n} + H_0\log k + K, \quad (5.24)$$

which corresponds to the indefinite integral of the Bosonic generating function, because it satisfies $\mathrm{d}\varphi(k)/\mathrm{d}k = \sum_{n\in\mathbb{Z}}H_n k^{-n-1}$. The 'constant of integration' K is not itself well defined as an operator, but if we fix the commutation relations

$$[H_0, K] = 1 \quad \text{and} \quad [K, H_n] = 0 \quad \text{for } n \neq 0, \quad (5.25)$$

then one can formally deduce the commutation relations that e^K in the above definition must have with H_n. If moreover we insist that K is a creation operator and H_0 an annihilation operator, then (5.23) can be written in the mnemonic form

$$\Psi(k) = \;:\mathrm{e}^{\varphi(k)}:_B, \quad \Psi^*(k) = \;:\mathrm{e}^{-\varphi(k)}:_B.$$

After the above preliminaries, our problem has the following answer:

Theorem 5.2 (Boson–Fermion correspondence) *The Fermionic generating functions $\psi(k)$ and $\psi^*(k)$ are realised in the Bosonic Fock space by (5.23). That is, for any $|u\rangle \in \mathcal{F}$ we have*

$$\Phi(\psi(k)|u\rangle) = \Psi(k)\Phi(|u\rangle) \quad \text{and} \quad \Phi(\psi^*(k)|u\rangle) = \Psi^*(k)\Phi(|u\rangle).$$

Proof The two proofs are exactly the same, so we prove the statement for $\psi(k)$ only. By definition, we have

$$\Phi(\psi(k)|u\rangle) = \sum_l z^l \langle l|e^{H(\mathbf{X})}\psi(k)|u\rangle = e^{\xi(\mathbf{X},k)} \sum_l z^l \langle l|\psi(k)e^{H(\mathbf{X})}|u\rangle.$$

On the other hand, if we write

$$\varepsilon(k^{-1}) = \left(\frac{1}{k}, \frac{1}{2k^2}, \frac{1}{3k^3}, \cdots\right)$$

then $e^{-\xi(\widetilde{\partial},k^{-1})}$ acts by translating the variables $f(\mathbf{x}) \mapsto f(\mathbf{x} - \varepsilon(k^{-1}))$, so that the statement reduces to the following lemma.

Lemma 5.3

$$\langle l|\psi(k) = k^{l-1}\langle l-1|e^{-H(\varepsilon(k^{-1}))}, \qquad (5.26)$$
$$\langle l|\psi^*(k) = k^{-l-1}\langle l+1|e^{H(\varepsilon(k^{-1}))}. \qquad (5.27)$$

Proof We suppose for simplicity that $l = 0$. The general case is similar. To prove (5.26), it is enough to show that

$$\langle \psi_{1/2} k^{-1} e^{-H(\varepsilon(k^{-1}))} \psi(p_1) \cdots \psi(p_{n-1}) \psi^*(q_n) \cdots \psi^*(q_1) \rangle$$
$$= \langle \psi(k)\psi(p_1) \cdots \psi(p_{n-1}) \psi^*(q_n) \cdots \psi^*(q_1) \rangle \qquad (5.28)$$

for any n. Here we note that

$$e^{\pm\xi(\varepsilon(k^{-1}),p)} = (1 - p/k)^{\mp 1},$$

so that taking $e^{-H(\varepsilon(k^{-1}))}$ over to the right-hand side, the left-hand side of (5.28) can be written

$$\frac{\prod_{i=1}^{n-1}(k-p_i)}{\prod_{j=1}^{n}(k-q_j)} \times \oint \frac{dk}{2\pi i} \langle \psi(k)\psi(p_1) \cdots \psi(p_{n-1}) \psi^*(q_n) \cdots \psi^*(q_1) \rangle.$$

Here the integral is obtained by expanding Cauchy's identity

$$\langle \psi(k)\psi(p_1) \cdots \psi(p_{n-1}) \psi^*(q_n) \cdots \psi^*(q_1) \rangle$$
$$= \frac{\prod_{i=1}^{n-1}(k-p_i)}{\prod_{j=1}^{n}(k-q_j)} \times \frac{\prod_{1\leq i<j\leq n-1}(p_i-p_j) \prod_{1\leq i<j\leq n}(q_j-q_i)}{\prod_{\substack{1\leq i\leq n-1 \\ 1\leq j\leq n}}(p_i-q_j)}$$

(see formula (5.4)) about $k = \infty$ and taking the coefficient of k^{-1}.

We apply the above formula to the right-hand side of (5.28), and the assertion follows easily. Q.E.D.

Exercises to Chapter 5

5.1. Prove relation (5.11).

5.2. Compute $\Phi(\psi_{-5/2}\psi^*_{-3/2}|\text{vac}\rangle)$.

5.3. If we consider x_n as weighted with weight n, and count an operator having energy d as having weight d, check that $H(\mathbf{x})$ of (5.13) is homogeneous of weight 0. Using this, together with the fact that $\langle l|$ has weight $l^2/2$, show that the polynomial $\Phi(|u\rangle)$ corresponding to $|u\rangle \in \mathcal{F}_l^{(d)}$ has weight $d - l^2/2$.

5.4. The generating function for the dimensions of the vector spaces $\mathcal{F}_l^{(d)}$

$$\text{ch}\mathcal{F} =_{\text{def}} \sum_{\substack{l \in \mathbb{Z} \\ d \geq l^2/2}} \dim(\mathcal{F}_l^{(d)}) z^l q^d$$

is called the *character* of the Fock space $\mathcal{F}$. Using the fact that (4.11) is a basis of $\mathcal{F}$, prove the formula

$$\text{ch}\mathcal{F} = \prod_{\substack{j > 0 \\ j \in \mathbb{Z}+1/2}} (1 + zq^j)(1 + z^{-1}q^j).$$

5.5. Show that on calculating the character using the Bosonic Fock space, we get

$$\text{ch}\mathcal{F} = \sum_{l \in \mathbb{Z}} z^l q^{l/2} \prod_{j=1}^{\infty} (1-q^j)^{-1}.$$

Use this together with Exercise 5.3 to verify the following formula (the Jacobi triple identity).

$$\prod_{j=1}^{\infty} (1-zq^{j-1})(1-z^{-1}q^j)(1-q^j) = \sum_{l \in \mathbb{Z}} (-z)^l q^{l(l-1)/2}.$$

Conversely, if we assume this identity is known, it follows that the map (5.16) is injective.

6
Transformation groups and tau functions

We start by showing that the space of all quadratic expressions in Fermions has a natural structure of an infinite dimensional Lie algebra. The rest of the chapter is taken up with a treatment of the group corresponding to this Lie algebra as a transformation group taking solutions of the KP equation into other solutions. To describe what happens in geometric language, this group action moves the vacuum vector around an orbit, each point of which is a tau function as in Chapter 2. In the huge infinite space of all functions, the orbit of this action is a submanifold, and its defining equation is nothing other than the Hirota equation.

6.1 Group actions and orbits

Given a space with a group acting on it, the locus traced out in the space by a point moving under the group action forms a certain figure. As you might expect, this figure acquires a high degree of symmetry from the very fact of the group action.

Quite generally, to say that a group G acts on a set S means that a point $gx \in S$ is specified for any element $g \in G$ of the group and every point $x \in S$ of the set, in such a way that the conditions $(g_1 g_2)x = g_1(g_2 x)$ and $ex = x$ hold (where $e \in G$ is the identity element). In this set-up, the subset $Gx = \{gx | g \in G\} \subset S$ is called the *orbit* of x under G. The central theme of this chapter is the orbit of an infinite dimensional group acting on an infinite dimensional space.

Example 6.1 *We fix the diagonal matrix*

$$J = \begin{pmatrix} 1 & & \\ & -1 & \\ & & -1 \end{pmatrix},$$

and consider the group
$$G = \{g \in M_{3\times 3}(\mathbb{R}) \mid {}^t g J g = J\}, \tag{6.1}$$
consisting of real 3×3 matrices preserving J. Suppose that $\vec{x}_0$ is the point
$$\vec{x}_0 = \begin{pmatrix} 1 \\ 0 \\ 0 \end{pmatrix};$$
what is the orbit of $\vec{x}_0$ under G? We see at once that $\vec{x}_0$ is a point of the hyperboloid of two sheets given by
$$ {}^t\vec{x} J \vec{x} \equiv x^2 - y^2 - z^2 = 1, \quad \text{where} \quad \vec{x} = \begin{pmatrix} x \\ y \\ z \end{pmatrix}. \tag{6.2}$$
On the other hand, by the very definition of G, for $g \in G$ we have
$${}^t(g\vec{x}_0) J (g\vec{x}_0) = {}^t\vec{x}_0 {}^t g J g \vec{x}_0 = {}^t\vec{x}_0 J \vec{x}_0 = 1.$$
It is not hard to see that the orbit $G\vec{x}_0$ of x_0 is the hyperboloid of two sheets (6.2) (see Exercise 6.1).

6.2 The Lie algebra $\mathfrak{gl}(\infty)$ of quadratic expressions

We write W for the set of all linear combinations of Fermions. The basic property of Fermions is that the anticommutator $[w, w']_+$ of any two elements $w, w' \in W$ is a number, that is, an element of $\mathbb{C}$. Now calculating the commutator (not the anticommutator) of quadratic expressions in Fermions gives
$$\begin{aligned}{}
[w_1 w_2, w_3 w_4] &= w_1 [w_2, w_3 w_4] + [w_1, w_3 w_4] w_2 \\
&= w_1 [w_2, w_3]_+ w_4 - w_1 w_3 [w_2, w_4]_+ \\
&\quad + [w_1, w_3]_+ w_4 w_2 - w_3 [w_1, w_4]_+ w_2.
\end{aligned}$$
The right-hand side is again a quadratic expressions in Fermions. We thus conclude that the vector subspace
$$W^{(2)} = \left\{ \sum_{i,j} w_i w_j \mid w_i \in W \right\} \subset \mathcal{A}$$
of quadratic expressions is closed under commutator brackets, and therefore forms a Lie algebra. We observe that, in particular, $W^{(2)}$ contains $\mathbb{C}$ as the set of anticommutators $[w, w']_+$.

6.2 The Lie algebra $\mathfrak{gl}(\infty)$ of quadratic expressions

Now we consider especially the subalgebra in $W^{(2)}$ formed by elements of charge 0. Any such element can be written in a unique way in the form

$$\sum_{m,n \in \mathbb{Z}+1/2} a_{mn} \psi_{-m} \psi_n^* + a_0, \quad \text{with } a_{mn}, a_0 \in \mathbb{C}. \tag{6.3}$$

Carrying out the computation as above, we get

$$[\psi_{-m}\psi_n^*, \psi_{-m'}\psi_{n'}^*] = \delta_{nm'}\psi_{-m}\psi_{n'}^* - \delta_{n'm}\psi_{-m'}\psi_n^* \tag{6.4}$$

for $m, n \in \mathbb{Z} + \frac{1}{2}$. We compare these with the elementary matrices

$$E_{mn} = (\delta_{im}\delta_{jn})_{i,j \in \mathbb{Z}+1/2};$$

that is, E_{mn} has 1 in the (m,n)th position and 0 everywhere else. Ordinary matrix multiplication gives $E_{mn}E_{m'n'} = \delta_{nm'}E_{mn'}$, so that (6.4) says that the elements $\psi_{-m}\psi_n^*$ have exactly the same commutation relations as the elementary matrices E_{mn}.

We now want to generalise (6.3) to a type of infinite sum which includes as a particular case the H_n introduced in (5.7). As we saw in Chapter 5, to handle this type of infinite sum, the Fermions must be recast as normal products. Thus, for an infinite matrix $A = (a_{mn})_{m,n \in \mathbb{Z}+1/2}$, consider the Fermion of (6.3) with the products replaced by normal products:

$$X_A = \sum_{m,n} a_{mn} : \psi_{-m}\psi_n^* : . \tag{6.5}$$

Now we study what effect the modification (6.5) has on the commutation relations. First, suppose that (6.5) is a finite sum. Noting that the normal product $: \psi_{-m}\psi_n^* :$ differs from $\psi_{-m}\psi_n^*$ only by a constant term, we get

$$\begin{aligned}
[X_A, X_B] &= \sum a_{ij}b_{kl}[\psi_{-i}\psi_j^*, \psi_{-k}\psi_l^*] \\
&= \sum a_{ij}b_{kl}\left(\delta_{jk}\psi_{-i}\psi_l^* - \delta_{li}\psi_{-k}\psi_j^*\right) \\
&= \sum a_{ij}b_{jl}\left(: \psi_{-i}\psi_l^* : + \delta_{il}\theta(i<0)\right) \\
&\quad - \sum b_{ki}a_{ij}\left(: \psi_{-k}\psi_j^* : + \delta_{kj}\theta(j<0)\right) \\
&= X_{[A,B]} + \omega(A,B),
\end{aligned} \tag{6.6}$$

where

$$\omega(A, B) = \sum a_{ij}b_{ji}(\theta(i<0) - \theta(j<0)) = -\omega(B, A), \tag{6.7}$$

and θ is the Boolean characteristic function as explained after (4.18). In

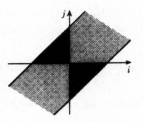

Fig. 6.1. The finiteness condition (6.8)

other words, compared to the matrix commutator $[A, B]$, only the constant term $\omega(A, B)$ has changed. Now we impose the following condition on the matrix $A = (a_{ij})$:

there exists $N > 0$ such that $a_{ij} = 0$ for all i, j with $|i - j| > N$. (6.8)

See Figure 6.1; for example, the sum in (6.7) runs over the black regions.

For example, the matrix of coefficients of H_n of (5.7) is $a_{ij} = \delta_{i+n,j}$, and satisfies (6.8). Provided that this condition holds, even for infinite matrices, the matrix commutator and the calculations for (6.7) both involve only finite sums. In the same way as for the H_n in Chapter 5, for any vector $|u\rangle$ of the Fock space, assuming (6.8), it is easy to see that $X_A |u\rangle$ has only finitely many nonzero terms.

Definition 6.1 *The Lie algebra* $\mathfrak{gl}(\infty)$ *is defined to be the vector space*

$$\mathfrak{gl}(\infty) = \{X_A \mid A \text{ satisfies (6.8)}\} \oplus \mathbb{C}, \qquad (6.9)$$

with the commutator bracket defined by (6.6).

The thing that ensures that the commutator bracket $[-, -]$ defined in (6.6) satisfies the Jacobi identity is the following *cocycle condition* on the constant correction term $\omega(A, B)$ of (6.7):

$$\omega(A, [B, C]) + \omega(B, [C, A]) + \omega(C, [A, B]) = 0. \qquad (6.10)$$

(See Exercise 6.2.)

At the same time, we note that taking the commutator with an element $X_A \in \mathfrak{gl}(\infty)$ induces a linear transformation on Fermions:

$$[X_A, \psi_{-n}] = \sum_m a_{mn} \psi_{-m}, \quad [X_A, \psi_n^*] = \sum_m (-a_{nm}) \psi_m^* \qquad (6.11)$$

(the right-hand sides are both finite sums). Here the transformation matrices corresponding to the ψ and the ψ^* are exactly the contragredient

6.2 The Lie algebra $\mathfrak{gl}(\infty)$ of quadratic expressions

of one another (that is, in the relation of M and $-{}^t M$), so that we have the relation

$$\sum_{n \in \mathbb{Z}+1/2} \left(\langle u|X_A \psi_{-n}|v\rangle \langle u'|\psi_n^*|v'\rangle + \langle u|\psi_{-n}|v\rangle \langle u'|X_A \psi_n^*|v'\rangle \right)$$
$$= \sum_{n \in \mathbb{Z}+1/2} \left(\langle u|\psi_{-n} X_A|v\rangle \langle u'|\psi_n^*|v'\rangle + \langle u|\psi_{-n}|v\rangle \langle u'|\psi_n^* X_A|v'\rangle \right) \quad (6.12)$$

for any $\langle u|, \langle u'|, |v\rangle, |v'\rangle$. We will make use of this fact later.

The elements of $\mathfrak{gl}(\infty)$ have charge 0, so their action on $\mathcal{F}$ takes each $\mathcal{F}_l$ to itself. In other words, $\mathcal{F}_l$ is a representation space of the Lie algebra $\mathfrak{gl}(\infty)$. We now use the Boson–Fermion correspondence to interpret these representations in the space of (polynomial) functions of $\mathbf{x}$. For this, it is enough to use the representation of Theorem 5.2 in terms of the generating functions and to consider the product $\psi(p)\psi^*(p)$. In the equality

$$\Psi(p)\Psi^*(q) = E(p)E^*(q)e^{K} p^{H_0} e^{-K} q^{-H_0},$$

we have

$$e^{-\xi(\tilde{\partial}, p^{-1})} e^{-\xi(\mathbf{x}, q)} = \frac{1}{1 - q/p} e^{-\xi(\mathbf{x}, q)} e^{-\xi(\tilde{\partial}, p^{-1})},$$

and using $p^{H_0} e^{-K} = p^{-1} e^{-K} p^{H_0}$ we arrive at the following expression:

$$\Psi(p)\Psi^*(q) = \frac{1}{p-q} e^{\xi(\mathbf{x},p) - \xi(\mathbf{x},q)} e^{-\xi(\tilde{\partial}, p^{-1}) + \xi(\tilde{\partial}, q^{-1})} p^{H_0} q^{-H_0}.$$

In particular, when we restrict this to the charge $l = 0$ subspace, $p^{H_0} q^{-H_0}$ acts as the identity, so that we can forget it. Now bearing in mind (5.3) and (5.2), we can express $:\psi(p)\psi^*(q):$ in terms of the vertex operator (3.17) of the KP equation

$$Z(p, q) = \frac{1}{p - q}(X(p, q) - 1). \quad (6.13)$$

We summarise our conclusion as follows.

Theorem 6.1 (Vertex operator representation of $\mathfrak{gl}(\infty)$) *We suppose
that the operators Z_{ij} acting on Bosonic Fock space $\mathbb{C}[x_1, x_2, x_3, \ldots]$ are defined by means of the generating function*

$$Z(p, q) = \sum_{i,j \in \mathbb{Z}+1/2} Z_{ij} p^{-i-1/2} q^{-j-1/2}. \quad (6.14)$$

Then
$$\sum_{m,n} a_{mn} :\psi_m \psi_n^*: \;\mapsto\; \sum_{m,n} a_{mn} Z_{-mn}$$

determines a representation of the Lie algebra $\mathfrak{gl}(\infty)$ on $\mathbb{C}[x_1, x_2, x_3, \ldots]$.

6.3 The transformation group of the KP hierarchy

We now gradually set out to explain how we can use the Boson–Fermion correspondence to obtain a unified construction of both the system of linear equations defining the KP hierarchy and the tau functions.

We define the group $\mathbf{G}$ corresponding to the Lie algebra $\mathfrak{gl}(\infty)$ to be

$$\mathbf{G} =_{\text{def}} \{ e^{X_1} e^{X_2} \cdots e^{X_k} \mid X_i \in \mathfrak{gl}(\infty) \}. \tag{6.15}$$

Here the question of whether the e^X are meaningful deserves closer scrutiny, but we do not go into the arguments at present. The object we want to consider is the orbit of the vacuum state under the action of $\mathbf{G}$:

$$\mathbf{G}|\text{vac}\rangle = \{ g|\text{vac}\rangle \mid g \in \mathbf{G} \}. \tag{6.16}$$

By the Boson–Fermion correspondence, we can view a point of the orbit (6.16) as a function

$$\tau(\mathbf{x}) = \tau(\mathbf{x};g) = \langle\text{vac}|e^{H(\mathbf{x})}g|\text{vac}\rangle. \tag{6.17}$$

A function of this form is called a *tau function*. Together with this, for any $g \in \mathbf{G}$, we introduce the *wave function* $w(\mathbf{x},k)$ and the *dual wave function* $w^*(\mathbf{x},k)$ by the following formulas:

$$w(\mathbf{x},k) = \frac{\langle 1|e^{H(\mathbf{x})}\psi(k)g|\text{vac}\rangle}{\langle\text{vac}|e^{H(\mathbf{x})}g|\text{vac}\rangle}; \tag{6.18}$$

$$w^*(\mathbf{x},k) = \frac{\langle -1|e^{H(\mathbf{x})}\psi^*(k)g|\text{vac}\rangle}{\langle\text{vac}|e^{H(\mathbf{x})}g|\text{vac}\rangle}. \tag{6.19}$$

In view of (5.19), and bearing in mind the relations (5.26)–(5.27), we see that these functions can be identified with the tau functions constructed in Chapter 3, (3.21)–(3.22). We show that we can derive the bilinear identity again starting from the expressions (6.17)–(6.19).

Theorem 6.2

$$\text{Res}_{k=\infty}\bigl(w^*(\mathbf{x},k)w(\mathbf{x}',k)\bigr) = 0 \quad \text{for all } \mathbf{x},\mathbf{x}'. \tag{6.20}$$

6.3 The transformation group of the KP hierarchy

Proof First, note that if $g \in \mathbf{G}$ then

$$\sum_{n \in \mathbb{Z}+1/2} \langle u|g\psi_{-n}|v\rangle \langle u'|g\psi_n^*|v'\rangle = \sum_{n \in \mathbb{Z}+1/2} \langle u|\psi_{-n}g|v\rangle \langle u'|\psi_n^*g|v'\rangle.$$

In fact for $g = e^{X_A}$, this formula is a consequence of its infinitesimal form (6.11), and for the general case it is enough to apply this repeatedly. Now for any n, one of ψ_{-n} and ψ_n^* must be an annihilation operator, so that using the above formula we get

$$-\operatorname{Res}_{k=\infty} \bigl(\langle u|\psi^*(k)g|\mathrm{vac}\rangle \langle u'|\psi(k)g|\mathrm{vac}\rangle \bigr)$$
$$= \sum_{n \in \mathbb{Z}+1/2} \langle u|\psi_n^* g|\mathrm{vac}\rangle \langle u'|\psi_{-n} g|\mathrm{vac}\rangle$$
$$= \sum_{n \in \mathbb{Z}+1/2} \langle u|g\psi_n^*|\mathrm{vac}\rangle \langle u'|g\psi_{-n}|\mathrm{vac}\rangle$$
$$= 0.$$

Taking $\langle u| = \langle 1|e^{H(\mathbf{x})}$ and $\langle u'| = \langle -1|e^{H(\mathbf{x}')}$, we arrive at the required formula. Q.E.D.

Example 6.2 *We calculate one example of a polynomial solution of the KP equation. Let a, b be constants, and set $X = a\psi_{-1/2}\psi^*_{-3/2} + b\psi_{-3/2}\psi^*_{-1/2}$. The corresponding group element is*

$$g = e^X = 1 + a\psi_{-1/2}\psi^*_{-3/2} + b\psi_{-3/2}\psi^*_{-1/2} + ab\psi_{-1/2}\psi^*_{-3/2}\psi_{-3/2}\psi^*_{-1/2}.$$

Then we get

$$\langle \psi_{-1/2}(\mathbf{x})\psi^*_{-3/2}(\mathbf{x}) \rangle = x_2 - \frac{1}{2}x_1^2, \quad \langle \psi_{-3/2}(\mathbf{x})\psi^*_{-1/2}(\mathbf{x}) \rangle = x_2 + \frac{1}{2}x_1^2,$$

$$\langle \psi_{-1/2}(\mathbf{x})\psi^*_{-3/2}(\mathbf{x})\psi_{-3/2}(\mathbf{x})\psi^*_{-1/2}(\mathbf{x}) \rangle = -x_1 x_3 + x_2^2 + \frac{x_1^4}{12}. \quad (6.21)$$

(The last expression is calculated using Wick's theorem.) Therefore the required tau function is

$$\tau(\mathbf{x}; g) = 1 + a\left(x_2 - \frac{1}{2}x_1^2\right) + b\left(x_2 + \frac{1}{2}x_1^2\right) + ab\left(-x_1 x_3 + x_2^2 + \frac{x_1^4}{12}\right).$$

Moreover, a, b are arbitrary, so that in particular, as $a, b \to \infty$, (6.21) itself is a solution of the KP equation.

As we have already shown at the end of Chapter 3 it follows from (6.20) that $w(\mathbf{x}, k)$ must satisfy a series of equations in Lax form with respect to the variables $x_1, x_2, x_3, \ldots$. Moreover, we have already seen that, from the equivalent expression (3.20), we can derive directly the hierarchy of differential equations for the tau function in Hirota form. In fact conversely, it can also be shown (see Chapter 9) that a polynomial $\tau(\mathbf{x})$ satisfying the bilinear identity can necessarily be written in the form (6.17) for some $g \in \mathbf{G}$. Therefore, finally, the tau functions are really the points of the orbits of $|\text{vac}\rangle$ under $\mathbf{G}$, and whether you write it as a Hirota differential equation, or the form of linear equation corresponding to wave functions, the KP hierarchy is really nothing other than the equations characterising the orbit. Putting this all together, we obtain the following picture.

Fermionic picture :   Fock space $\supset \mathbf{G}|\text{vac}\rangle$.

Bosonic picture : $\mathbb{C}[x_1, x_2, x_3, \ldots] \supset \{\text{tau functions}\}$.

By definition $\mathbf{G}$ acts naturally on the orbits. In constructing soliton solutions, the vertex operators (6.14) introduced out of the blue are in Fermionic terms just the quadratic expressions which are the generators of the Lie algebra, and act on the orbit as infinitesimal transformations. In other words, the vertex operators give infinitesimal transformations of solutions of the KP hierarchy (tau functions)

Exercises to Chapter 6

6.1. Verify that (6.1) defines a group G. Show moreover that the orbit of $\vec{x}_0$ is determined by (6.2).

6.2. Verify the cocycle condition (6.10) for any A, B, C satisfying (6.8).

6.3. Derive (6.21) and verify directly that substituting it in the KP equation (3.13) gives 0.

7
The transformation group of the KdV equation

Coming down from the abstract heights of the KP hierarchy, we return to the particular case of the KdV hierarchy. The world of solutions is then narrower, and the corresponding transformation group is also smaller. We give an introduction to the affine Lie algebra $\widehat{sl}_2$, which appears as the infinitesimal transformations of the KdV hierarchy.

7.1 KP hierarchy versus KdV hierarchy

The KP and KdV hierarchies were studied in Chapters 2 and 3. Here we want to compare them from various points of view and to summarise their properties. Both were introduced as the nonlinear systems of partial differential equations obtained as the compatibility conditions on systems of linear differential equations of the form

$$\frac{\partial w}{\partial x_n} = (L^n)_+ w, \quad Lw = kw, \quad \text{where } L = \partial + f_1 \partial^{-1} + f_2 \partial^{-2} + \cdots. \tag{7.1}$$

Both can be written as Hirota equations, and have n-soliton solutions for any n. For example, a 2-soliton solution is given by

$$\tau = 1 + c_1 e^{\xi_1} + c_2 e^{\xi_2} + c_1 c_2 a_{12} e^{\xi_1 + \xi_2}.$$

For more details, see Table 7.1.

The KdV hierarchy can be obtained as a specialisation of the KP hierarchy. In the language of pseudodifferential operators, as explained in Chapter 2, the specialisation condition is that L^2 should be a differential operator. It is clear from the system of linear equations (7.1) that in this case, in the power series expansion (3.23) of w, the coefficients w_l do not depend on the even-numbered time variables $x_2, x_4, x_6, \ldots$.

Table 7.1. *The KP hierarchy versus the KdV hierarchy*

	KP hierarchy	KdV hierarchy
time variables	$x_1, x_2, x_3, \ldots$	$x_1, x_3, x_5, \ldots$
pseudodiff. op.	$L = \partial + f_1 \partial^{-1} + f_2 \partial^{-2} + \cdots$	$L = (\partial^2 + u)^{1/2}$
nonlinear equation	$\dfrac{3}{4}\dfrac{\partial^2 u}{\partial x_2^2} = \dfrac{\partial}{\partial x}\left(\dfrac{\partial u}{\partial x_3} - \dfrac{3}{2}u\dfrac{\partial u}{\partial x} - \dfrac{1}{4}\dfrac{\partial^3 u}{\partial x^3}\right),$ $\ldots$	$\dfrac{\partial u}{\partial x_3} = \dfrac{3}{2}u\dfrac{\partial u}{\partial x} + \dfrac{1}{4}\dfrac{\partial^3 u}{\partial x^3},$ $\ldots$
Hirota equation	$(D_1^4 + 3D_2^2 - 4D_1 D_3)\tau \cdot \tau = 0,$ $\ldots$	$(D_1^4 - 4D_1 D_3)\tau \cdot \tau = 0,$ $\ldots$
solitons	$\xi_j = \sum_{n=1}^{\infty}(p_j^n - q_j^n)x_n$ $c_{ij} = \dfrac{(p_i - p_j)(q_i - q_j)}{(p_i - q_j)(q_i - p_j)}$	$\xi_j = 2\displaystyle\sum_{n=1,3,5,\ldots} p_j^n x_n$ $c_{ij} = \left(\dfrac{p_i - p_j}{p_i - p_j}\right)^2$

Moreover, the tau function τ is determined by the w_l from the relation (3.21). Therefore the tau functions of the KdV hierarchy are exactly obtained by imposing the additional condition

$$\frac{\partial \tau}{\partial x_n} = 0 \quad \text{for } n = 2, 4, 6, \ldots \tag{7.2}$$

on those of the KP hierarchy. For example, if we impose the relations

$$q_j = -p_j \tag{7.3}$$

on the parameters p_j, q_j in the soliton solutions of the KP hierarchy, then $x_2, x_4, x_6, \ldots$ drop out automatically, and we obtain soliton solutions of the KdV hierarchy (see Table 7.1).

Since solutions of the KdV hierarchy are particular cases of solutions of the KP hierarchy, we can expect that the transformation group taking solutions of the KdV hierarchy into themselves should be a subgroup of the corresponding group for the KP hierarchy. We now study this.

7.2 The Transformation group of the KdV equation

In view of the condition (7.3) on the soliton solutions, we consider the parameters p, q in the vertex operators (6.14), and impose the relation $p^2 = q^2$ on them. Corresponding to the two possibilities $q = \pm p$, we get

$$\left.\begin{aligned} Z(p,p) &= \sum_{n \in \mathbb{Z}} H_n p^{-n-1}, \\ Z(p,-p) &= \frac{1}{2p}\left(\exp\left(\sum_{\substack{n>0 \\ n \text{ odd}}} 2x_n p^n\right) \exp\left(-\sum_{\substack{n>0 \\ n \text{ odd}}} \frac{2}{n}\frac{\partial}{\partial x_n} p^{-n}\right) - 1\right). \end{aligned}\right\} \tag{7.4}$$

The first set are exactly the Bosons, the second set the vertex operators of the KdV equation introduced in Chapter 2. Except for the x_n with n even, the first set of operators all preserve (7.2), and therefore provide infinitesimal transformations of the KdV hierarchy. Now, what do these look like as a Lie algebra?

To rephrase (7.4) in terms of Fermions, we are restricting the framework from quite general quadratic expressions in the Fermions to the elements of the form

$$:\psi(p)\psi^*(p): \quad \text{and} \quad :\psi(p)\psi^*(-p): . \tag{7.5}$$

It is easy to see that this is equivalent to the following: instead of general linear combinations $X_A = \sum a_{mn} :\psi_{-n}\psi_n^*:$, we restrict attention only to the elements which are invariant under the translation

$$\psi_n \to \psi_{n-2}, \quad \psi_n^* \to \psi_{n+2}^*$$

of the index. This amounts simply to imposing the condition

$$a_{mn} = a_{m+2,n+2} \quad \text{for all } m, n \tag{7.6}$$

on the entries of A. Now an infinite matrix $A = (a_{mn})$ satisfying (7.6) is clearly determined by its entries a_{mn} with $m = \pm 1/2$ and $n \in \mathbb{Z} + 1/2$. It is convenient to express this as a Laurent polynomial in one variable t (that is, a polynomial in t, t^{-1}) in the following form:

$$A(t) = \sum_{j \in \mathbb{Z}} \begin{pmatrix} a_{-\frac{1}{2},-\frac{1}{2}+j} & a_{-\frac{1}{2},\frac{1}{2}+j} \\ a_{\frac{1}{2},-\frac{1}{2}+j} & a_{\frac{1}{2},\frac{1}{2}+j} \end{pmatrix} t^j.$$

In terms of this notation, (6.7) can be rewritten as follows:

$$\omega(A, B) = \text{Res}_{t=0} \left(\text{Tr} \left(\frac{dA}{dt}(t) B(t) \right) \right). \tag{7.7}$$

(See Exercise 7.1). The Lie algebra that arises in this way has the following name.

Definition 7.1 (Affine Lie algebra $\widehat{\mathfrak{sl}_2}$) *Consider the space generated by matrices whose entries are Laurent polynomials in t, together with one element c:*

$$\widehat{\mathfrak{sl}_2} = \left\{ \begin{pmatrix} \alpha(t) & \beta(t) \\ \gamma(t) & -\alpha(t) \end{pmatrix} \bigg| \alpha(t), \beta(t), \gamma(t) \in \mathbb{C}[t, t^{-1}] \right\} \oplus \mathbb{C} c.$$

This has a Lie algebra structure, with the commutator bracket:

$$[A(t), B(t)] = [A(t), B(t)]_{\text{mat}} + \text{Res}_{t=0} \left(\text{Tr} \left(\frac{dA}{dt}(t) B(t) \right) \right),$$

$$[c, X \otimes A(t)] = 0 \quad \text{for all } X.$$

In words, c commutes with everything. Here the subscript $[-, -]_{\text{mat}}$ means the commutator of matrices. This algebra is called the affine Lie algebra $\widehat{\mathfrak{sl}_2}$.

We summarise the story so far. Write $\mathcal{B}^{(2)}$ for the Heisenberg algebra, generated by the odd numbered variables $x_n, \partial/\partial x_n$ (for $n = 1, 3, 5, \ldots$). Acting by $\mathcal{B}^{(2)}$ on the vacuum state $1 \in \mathbb{C}[\mathbf{x}]$ gives the subspace

$$\mathbb{C}[x_1, x_3, x_5, \ldots]$$

as its orbit. The vertex operators (7.4) act on this space. Then $\mathcal{B}^{(2)}$

7.2 The Transformation group of the KdV equation

and the operators (7.4) give a representation of the affine Lie algebra $\widehat{sl}_2$. Here we suppose the central element c acts by 1. In general if c acts by a scalar k, we say the representation has *level k*. The infinitesimal transformations taking solutions of the KdV equations into themselves make up the affine Lie algebra $\widehat{sl}_2$ (in its the level 1 representation). The orbit of the vacuum state under $\widehat{sl}_2$, that is, the tau functions, give the solutions of the KdV equation (in Hirota form).

Remark 7.1 *Instead of the reduction considered in this section, we could more generally fix any natural number l and let $p^l = q^l$. This gives rise to a different series of soliton equations. The corresponding infinitesimal transformations is called $\widehat{sl}_l$, and gives rise to a series of Lie algebras. Affine Lie algebras are a class of infinite dimensional Lie algebras that are easy to handle, and are widely used in applications.*

Exercise to Chapter 7

7.1. Verify (7.7).

8
Finite dimensional Grassmannians and Plücker relations

The scene changes once more, and this chapter gives an introduction to Grassmann varieties. The link between this classical notion of projective geometry and the material of the book so far is provided by the Plücker relations. We explain the Plücker relations in the case of finite dimensional vector spaces as an introduction to the material of the following chapters.

8.1 Finite dimensional Grassmannians

The subject matter we now turn to arises from developments in projective geometry, or can be viewed as supplementary material in linear algebra. Starting our narrative in this way will not seem out of place in substance to a modern reader, although it might have provoked considerable conceptual resistance in the 19th century. One of the areas closely related to what we are about to explain would have been described by words like *line geometry*. This was part of a fertile area of research in 19th century mathematics, and it has many relations to current interests.

We fix an N dimensional vector space V and a natural number m with $0 \leq m \leq N$, and write $\text{Grass}(m, V)$ or $\text{Grass}(m, N)$ for the set of m dimensional vector subspaces of V. There are various types of them, depending on the choice of V, N and m, and we call them all collectively *Grassmann varieties* or *Grassmannians*.

For the case $m = 0$, there is only one 0 dimensional subspace $\{0\}$ of any vector space V, so that $\text{Grass}(0, V)$ consists of just one point $\{0\}$. For our next example, we consider the case $m = 1$. Then $\text{Grass}(1, V)$ is the set of 1 dimensional vector subspaces of V. In other words, an element of $\text{Grass}(1, V)$ is determined by a nonzero vector $\mathbf{v} \in V$, with two vectors $\mathbf{v}, \mathbf{v}'$ describing the same point of $\text{Grass}(1, V)$ if and only if they are linearly dependent, in other words, scalar multiples of one another.

8.1 Finite dimensional Grassmannians

Grass$(1, V)$ is usually called *complex projective space*, and denoted by $\mathbb{P}(V)$ or $\mathbb{P}_{N-1}(\mathbb{C})$. (Here we are taking the coefficient field to be the field $\mathbb{C}$ of complex numbers.)

We consider in particular the case $N = 2$. In this case, as we discuss presently, $\mathbb{P}_1(\mathbb{C})$ is 1 dimensional, and is called the *complex projective line*. A point of $\mathbb{P}_1(\mathbb{C})$ is determined by a vector $\mathbf{v} = (v_1, v_2)$ with $v_1, v_2 \in \mathbb{C}$. The vectors with $v_2 = 0$ determine a point of $\mathbb{P}_1(\mathbb{C})$: there are lots of the vectors themselves, a whole line of them, but, according to the definition of projective space, they all determine the same point, called the *point at infinity* of $\mathbb{P}_1(\mathbb{C})$. To a vector $\mathbf{v}$ with $v_2 \neq 0$, we assign the number v_1/v_2. This gives a one-to-one correspondence between $\mathbb{P}_1(\mathbb{C})$ minus the point at infinity and the complex plane. Now if you delete one point from the 2 dimensional sphere, you get the plane (think of the sphere as being made of rubber; see Exercise 8.1).

Although this argument was fairly rough, we see that $\mathbb{P}_1(\mathbb{C})$ can be identified with the 2 dimensional sphere. A slightly more detailed argument can be used to prove that $\mathbb{P}_1(\mathbb{C})$ is homeomorphic to the 2 dimensional sphere, in fact even isomorphic as a complex manifold. In the same way, $\mathbb{P}_{N-1}(\mathbb{C})$ contains a copy of $\mathbb{C}^{N-1}$, although now $\mathbb{P}_{N-1}(\mathbb{C}) \setminus \mathbb{C}^{N-1}$ is more than just one point.

The projective space $\mathbb{P}(V)$, and the Grassmannians Grass(m, V) generally, have the structure of an algebraic varieties. We refer elsewhere for an explanation of algebraic varieties (see for example I.R. Shafarevich, *Basic algebraic geometry*, Springer, Chapter 1), but to get to grips with some of the ideas, we calculate here the dimension of the Grassmannian. The dimension of a variety is the number of continuous parameters required to specify a point of the variety; you can get a rough idea of dimension by analogy with that of a vector space, or convince yourself by calling to mind a few examples, such as the spheres (although here we usually have in mind dimension over $\mathbb{C}$). In the case of projective space, we specify a point of $\mathbb{P}(V)$ by fixing one nonzero vector, up to the indeterminacy of a scalar multiple; therefore $\mathbb{P}(V)$ is $N - 1$ dimensional.

For general values of m, we argue as follows: pick a basis of V, that is, a set of N linearly independent vectors $(\mathbf{v}_1, \ldots, \mathbf{v}_N)$. Then we can identify the elements of V with row vectors of length N. An m dimensional vector subspace $W \subset V$ is determined by m linearly independent vectors $\mathbf{w}_1, \ldots, \mathbf{w}_m$. We can then set

$$\mathbf{w}_i = \sum_{j=1}^{N} v_{ij} \mathbf{v}_j. \tag{8.1}$$

68   *Finite dimensional Grassmannians and Plücker relations*

In other words, we specify an m dimensional vector subspace $W \subset V$ by writing down an $m \times N$ matrix $M = (v_{ij})$ of rank m. The matrix $M = M_W$ is called a *frame* of the subspace W. However, there is some waste in specifying W by means of a frame. A necessary and sufficient condition for two frames M_1 and M_2 to describe the same subspace W is that one can be obtained from the other by change of basis, that is $M_2 = gM_1$ with $g \in \mathrm{GL}(m, \mathbb{C})$. Our present purpose is only to calculate the dimension of $\mathrm{Grass}(m, V)$, so that we can change the order of the basis of V if necessary, and suppose that M is of the form

$$\begin{pmatrix} 1 & & & 0 & * & * & * \\ & 1 & & & * & * & * \\ & & \ddots & & \vdots & \vdots & \vdots \\ & 0 & & & * & * & * \\ & & & 1 & * & * & * \end{pmatrix},$$

where the $m(N - m)$ entries of the right-hand block are arbitrary. It follows from this that the dimension of $\mathrm{Grass}(m, V)$ is $m(N - m)$.

Here is an alternative way of seeing this. Suppose that we choose a basis of W, and extend it to a basis of V. The freedom in the choice of basis of V is described by $\mathrm{GL}(N, \mathbb{C})$. However, of these, the set of matrices which leave the subspace W invariant (as a subset) is given by the subset of matrices of the block upper diagonal form:

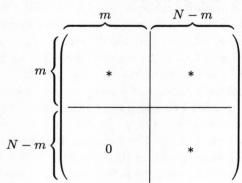

(The set of all matrices of this form is a subgroup of $\mathrm{GL}(N, \mathbb{C})$, called a *parabolic subgroup*.) It follows from this that $\mathrm{Grass}(m, V)$ has

$$N^2 - m^2 - (N - m)^2 - m(N - m) = m(N - m)$$

degrees of freedom, that is, dimension $m(N - m)$.

After projective space itself, the simplest case of a Grassmannian is when $N = 4$ and $m = 2$. This variety Grass(2, 4) has an embedding in projective space $\mathbb{P}_5(\mathbb{C})$ discovered by Plücker and studied in detail by Klein and others (see Example 8.1).

8.2 Plücker coordinates

As we said in the preceding section, the Grassmannian Grass(m, V) has the structure of an algebraic variety. The formal meaning of this assertion is that the points of Grass(m, V) have coordinates which are common solutions of a system of algebraic equations, and that all the solutions of this system are given by points of Grass(m, V). As an example, a circle is an algebraic variety in this sense: a point of the plane has coordinates (x, y), and the circle is the set of all points whose coordinates satisfy the equation

$$x^2 + y^2 - r^2 = 0 \quad \text{for some fixed } r > 0.$$

Here we should note the following: if we take an algebraic variety in the above simple minded sense, then we have first of all to specify the space where our variety is contained, and how it is contained there; there may be lots of possible ways of doing this. For example, when we consider the circle, there may be many cases in which we do not really care what space it is contained in. Although at this point we do not want to go into this problem in depth, we do need to consider the space in which the Grassmannian is contained.

We start again from the case $m = 1$. As we have already noted, a point of Grass($1, V$) = $\mathbb{P}(V)$ is described by a 1 dimensional vector subspace of V, spanned by a nonzero vector $\mathbf{v} = (v_1, v_2, \ldots, v_N) \in V$. We take the components of the vector $\mathbf{v} \in V$ as the coordinates of the point of $\mathbb{P}(V)$; they are called the *homogeneous coordinates* of $\mathbb{P}(V)$. These coordinates must satisfy the condition that the components are not all 0:

$$(v_1, \ldots, v_N) \neq (0, \ldots, 0).$$

The homogeneous coordinates do not correspond one to one to points of $\mathbb{P}(V)$: we note that a necessary and sufficient condition for two sets of coordinates $(v_1, v_2, \ldots, v_N)$ and $(v'_1, v'_2, \ldots, v'_N)$ to describe the same point is that there should exist a nonzero number c such that $v'_i = cv_i$ for all $i = 1, \ldots, N$. Thus we want to think of $\mathbb{P}(V)$ itself as being the space which contains points with this type of coordinates.

For general m, the space in which the Grassmannian Grass(m, V) is contained is usually taken to be a projective space of sufficiently large dimension. We now explain how to do this. Each point $W \in \mathrm{Grass}(m, V)$ of the Grassmannian is an m dimensional vector subspace $W \subset V$, and corresponds to a matrix $M_W = (v_{ij})$ with m rows and N columns. We want to consider all the $m \times m$ minors of M

$$v_\alpha = \det\left(v_{i,\alpha_j}\right)_{1 \leq i,j \leq m},$$

where $\alpha = (\alpha_1, \ldots, \alpha_m)$ with $1 \leq \alpha_1 < \cdots < \alpha_m \leq N$; we think of these as written out as a list of $\binom{N}{m}$ minors. If we change the basis of M, we multiply M on the left by a nondegenerate $m \times m$ matrix $h \in \mathrm{GL}(m, \mathbb{C})$. The effect of this on the v_α is to multiply them all simultaneously by $\det h$ (see Exercise 8.2).

There are $\binom{N}{m}$ different ways of choosing an $m \times m$ minor of M, so that we can view the v_α as homogeneous coordinates of projective space of dimension $\binom{N}{m-1}$. It can be seen that two different points $W, W' \in \mathrm{Grass}(m, V)$ correspond to different $(v_\alpha), (v'_\alpha)$. These coordinates (v_α) are called the *Plücker coordinates* of a point $W \in \mathrm{Grass}(m, V)$.

Here is an alternative way of thinking about the Plücker coordinates: we assume that the reader already knows about tensor products of vector spaces; if necessary, refer to a suitable textbook (for example S. Lang, *Algebra*, second edition, Addison–Wesley, 1984, Chap. XVI). Consider the mth exterior product $\bigwedge^m V$ of V. A necessary and sufficient condition for m vectors $\mathbf{w}_1, \ldots, \mathbf{w}_m \in V$ to be linearly independent is that $\mathbf{w}_1 \wedge \cdots \wedge \mathbf{w}_m \neq 0 \in \bigwedge^m V$. If $W \subset V$ is an m dimensional vector subspace, let $\mathbf{w}_1, \ldots, \mathbf{w}_m$ be a basis. From what we have said, we know that $\mathbf{w}_1 \wedge \cdots \wedge \mathbf{w}_m \neq 0$, so that W corresponds to a nonzero vector of $\bigwedge^m V$. However, this correspondence depends on the choice of basis. Let $\mathbf{w}'_1, \ldots, \mathbf{w}'_m$ be another basis, and h the matrix giving the change of basis:

$$\mathbf{w}'_i = \sum_{j=1}^m h_{ji} \mathbf{w}_j;$$

then

$$\mathbf{w}'_1 \wedge \cdots \wedge \mathbf{w}'_m = \det(h) \mathbf{w}_1 \wedge \cdots \wedge \mathbf{w}_m$$

(see Exercise 8.2). Thus changing the basis of W just multiplies the vector by a scalar. Thus we see that there is a well-defined map $\mathrm{Grass}(m, V) \to \mathbb{P}(\bigwedge^m V)$, called the *Plücker embedding*. The relation between Plücker coordinates and the notation (8.1) is given by

$$\mathbf{w}_1 \wedge \cdots \wedge \mathbf{w}_m = \sum_{\alpha_1 < \cdots < \alpha_m} v_{\alpha_1,\ldots,\alpha_m} \mathbf{v}_{\alpha_1} \wedge \cdots \wedge \mathbf{v}_{\alpha_m}.$$

From this point of view, insisting that the indices $\alpha_1, \ldots, \alpha_m$ on the Plücker coordinates $v_{\alpha_1,\ldots,\alpha_m}$ are ordered as $\alpha_1 < \cdots < \alpha_m$ may turn out to be inconvenient, and it is more natural to regard the Plücker coordinates as skew-symmetric in the indices. This is the point of view we adopt in what follows.

8.3 Plücker relations

The Plücker coordinate $(v_{\alpha_1,\ldots,\alpha_m})$ of a point $\mathrm{Grass}(m, V)$ is skew-symmetric in the indices. However, not every set of $\binom{N}{m}$ coordinates which change signs on interchanging the indices is the image of an element of $\mathrm{Grass}(m, V)$.

It is easy to see that there is no linear relation satisfied by the Plücker coordinates of all points of $\mathrm{Grass}(m, V)$ (see Exercise 8.4). However, there are quadratic relations between them, of the following form.

Theorem 8.1 *Let* $1 \leq \alpha_1, \ldots, \alpha_{m-1} \leq N$ *and* $1 \leq \beta_1, \ldots, \beta_{m+1} \leq N$ *be two sets of pairwise distinct indices. Then the following relation holds:*

$$\sum_{i=1}^{m+1} (-1)^{i-1} v_{\alpha_1,\ldots,\alpha_{m-1},\beta_i} v_{\beta_1,\ldots,\beta_{i-1},\beta_{i+1},\ldots,\beta_{m+1}} = 0. \qquad (8.2)$$

The Plücker relations are often used in the form

$$v_{\alpha_1,\ldots,\alpha_m} v_{\beta_1,\ldots,\beta_m} = \sum_{i=1}^{m} v_{\alpha_1,\ldots,\alpha_{s-1},\beta_i,\alpha_{s+1},\ldots,\alpha_m} v_{\beta_1,\ldots,\beta_{i-1},\alpha_s,\beta_{i+1},\ldots,\beta_m}$$

(8.3)

for any s. This formula can be obtained from (8.2) by renumbering the indices.

Proof Writing out the left-hand side of (8.2) in more detail gives

$$\sum_{i=1}^{m+1} (-1)^{i-1} \begin{vmatrix} v_{1\alpha_1} & \cdots & v_{1\alpha_{m-1}} & v_{1\beta_i} \\ \cdot & \cdots & \cdot & \cdot \\ v_{m\alpha_1} & \cdots & v_{m\alpha_{m-1}} & v_{m\beta_i} \end{vmatrix}$$

$$\times \begin{vmatrix} v_{1\beta_1} & \cdots & v_{1\beta_{i-1}} & v_{1\beta_{i+1}} & \cdots & v_{1\beta_{m+1}} \\ \cdot & \cdots & \cdot & \cdot & \cdots & \cdot \\ v_{m\beta_1} & \cdots & v_{m\beta_{i-1}} & v_{m\beta_{i+1}} & \cdots & v_{m\beta_{m+1}} \end{vmatrix}.$$

72   *Finite dimensional Grassmannians and Plücker relations*

Expanding the first determinant down the final column gives

$$\sum_{i=1}^{m+1} (-1)^{i-1} \sum_{j=1}^{m} (-1)^{m+j} v_{j\beta_i} \widetilde{v}_j$$

$$\times \begin{vmatrix} v_{1\beta_1} & \cdots & v_{1\beta_{i-1}} & v_{1\beta_{i+1}} & \cdots & v_{1\beta_{m+1}} \\ \cdot & \cdots & \cdot & \cdot & \cdots & \cdot \\ v_{m\beta_1} & \cdots & v_{m\beta_{i-1}} & v_{m\beta_{i+1}} & \cdots & v_{m\beta_{m+1}} \end{vmatrix},$$

where we have set

$$\widetilde{v}_j = \begin{vmatrix} v_{1\alpha_1} & \cdots & v_{1\alpha_{m-1}} \\ \cdot & \cdots & \cdot \\ v_{j-1\alpha_1} & \cdots & v_{j-1\alpha_{m-1}} \\ v_{j+1\alpha_1} & \cdots & v_{j+1\alpha_{m-1}} \\ \cdot & \cdots & \cdot \\ v_{m\alpha_1} & \cdots & v_{m\alpha_{m-1}} \end{vmatrix}.$$

Putting this together gives

$$\sum_{j=1}^{m} (-1)^j \widetilde{v}_j \begin{vmatrix} v_{1\beta_1} & \cdots & v_{1\beta_{i-1}} & v_{1\beta_i} & v_{1\beta_{i+1}} & \cdots & v_{1\beta_{m+1}} \\ \cdot & \cdots & \cdot & \cdot & \cdot & \cdots & \cdot \\ v_{m\beta_1} & \cdots & v_{m\beta_{i-1}} & v_{m\beta_i} & v_{m\beta_{i+1}} & \cdots & v_{m\beta_{m+1}} \\ v_{j\beta_1} & \cdots & v_{j\beta_{i-1}} & v_{j\beta_i} & v_{j\beta_{i+1}} & \cdots & v_{j\beta_{m+1}} \end{vmatrix}.$$

But all of these determinants have a repeated row, and are thus zero. Q.E.D.

These relations can also be proved using the Laplace expansion of a $2m \times 2m$ determinant (see Exercise 8.5).

Example 8.1 *We write out the Plücker relations in the case $N = 4$, $m = 2$. As already mentioned, after projective space itself, this is the simplest Grassmannian of all. In this case the linearly independent coordinates are*

$$v_{12}, v_{13}, v_{14}, v_{23}, v_{24}, v_{34},$$

and taking $\{1\}, \{2, 3, 4\}$ in (8.2) gives the Plücker relation

$$v_{12} v_{34} - v_{13} v_{24} + v_{14} v_{23} = 0.$$

In fact the converse of Theorem 8.1 also holds.

Theorem 8.2 *Any nonzero collection of numbers $(v_{\alpha_1,\ldots,\alpha_m})$ which is skew-symmetric under interchanging the indices and satisfies the Plücker relations (8.2) is the Plücker coordinate of a point of $\mathrm{Grass}(m, V)$, that is, of an m dimensional vector subspace of V.*

8.3 Plücker relations

Proof By assumption, $(v_{\alpha_1,\ldots,\alpha_m})$ is a nonzero vector. We now suppose that $v_{1\cdots m} \neq 0$. (We can reduce to this case by renumbering the basis of V if necessary.) Passing to a constant multiple, we can arrange that moreover,

$$v_{1\cdots m} = 1,$$

and we assume this in what follows. Set

$$w_{ij} = v_{1\cdots i-1,j,i+1\cdots m} \quad \text{for } i = 1,\ldots,m \text{ and } j = 1,\ldots,N.$$

Then by construction, we have

$$w_{ij} = \delta_{ij} \quad \text{for } j \leq m, \tag{8.4}$$

so that this determines m vectors $\mathbf{w}_i = (w_{i1},\ldots,w_{iN})$, for $i = 1,\ldots,m$, which span an m dimensional vector subspace $W \subset V$. In addition, we set

$$w_{\alpha_1,\ldots,\alpha_m} = \begin{vmatrix} w_{1\alpha_1} & \cdots & w_{1\alpha_m} \\ \cdot & \cdots & \cdot \\ w_{m\alpha_1} & \cdots & w_{m\alpha_m} \end{vmatrix}.$$

What we are required to prove is that

$$w_{\alpha_1,\ldots,\alpha_m} = v_{\alpha_1,\ldots,\alpha_m} \quad \text{for all } \alpha_1,\ldots,\alpha_m. \tag{8.5}$$

We pick one set $\alpha_1,\ldots,\alpha_m$ of pairwise distinct indices. Now let $\gamma_1,\ldots,\gamma_s$ be the elements among $\alpha_1,\ldots,\alpha_m$ which are $> m$, and $\beta_1,\ldots,\beta_s$ the elements of $1,\ldots,m$ missing from the list $\alpha_1,\ldots,\alpha_m$. We can permute $\alpha_1,\ldots,\alpha_m$ into the following order:

$$1,\ldots,\beta_1-1,\gamma_1,\beta_1+1,\ldots,\beta_s-1,\gamma_s,\beta_s+1,\ldots,m.$$

Since (8.4) holds by assumption, if we expand out the $m \times m$ determinant for $w_{\alpha_1,\ldots,\alpha_m}$ down the $m-s$ columns not containing γ_i we get

$$w_{\alpha_1,\ldots,\alpha_m} = \begin{vmatrix} w_{\beta_1\gamma_1} & \cdots & w_{\beta_1\gamma_s} \\ \cdot & \cdots & \cdot \\ w_{\beta_s\gamma_1} & \cdots & w_{\beta_s\gamma_s} \end{vmatrix}.$$

Now, we prove (8.5) by induction on s. It is obvious if $s = 0$ or 1. So suppose that (8.5) holds for all $s < t$. By the Plücker relations (8.3), we get

$$v_{\alpha_1,\ldots,\alpha_m} v_{1,\ldots,m} = \sum_{i=1}^{m} v_{\alpha_1,\ldots,\alpha_{r-1},i,\alpha_{r+1},\ldots,\alpha_m} v_{1,\ldots,i-1,\alpha_r,i+1,\alpha_r,i+1,\ldots,m}.$$

Here r is arbitrary. We choose an r for which $\alpha_r = \gamma_t$. Then for $i \neq \beta_1, \ldots, \beta_t$ we get

$$v_{\alpha_1,\ldots,\alpha_{r-1},i,\alpha_{r+1},\ldots,\alpha_m} = 0.$$

Indeed, the numbers from 1 to m missing from the set $\{\alpha_1, \ldots, \alpha_m\}$ are exactly $\beta_1, \ldots, \beta_t$. When $i = \beta_j$, among the indices of $v_{\alpha_1,\ldots,\alpha_{r-1},\beta_j,\alpha_{r+1},\ldots,\alpha_m}$, the number missing from 1 to m is $t-1$, so by the inductive assumption,

$$v_{\alpha_1,\ldots,\alpha_{r-1},\beta_j,\alpha_{r+1},\ldots,\alpha_m} = \begin{vmatrix} w_{1\alpha_1} & \cdots & w_{1\alpha_{r-1}} & w_{1\beta_j} & w_{1\alpha_{r+1}} & \cdots & w_{1\alpha_m} \\ \cdot & \cdots & \cdot & \cdot & \cdot & \cdots & \cdot \\ w_{m\alpha_1} & \cdots & w_{m\alpha_{r-1}} & w_{m\beta_j} & w_{m\alpha_{r+1}} & \cdots & w_{m\alpha_m} \end{vmatrix}.$$

Now by using the above argument repeatedly, we can rewrite the above determinant as

$$v_{\alpha_1,\ldots,\alpha_{r-1},\beta_j,\alpha_{r+1},\ldots,\alpha_m} = \begin{vmatrix} w_{\beta_1\gamma_1} & \cdots & w_{\beta_1\gamma_{t-1}} & w_{\beta_1\beta_j} \\ \cdot & \cdots & \cdot & \cdot \\ w_{\beta_{j-1}\gamma_1} & \cdots & w_{\beta_{j-1}\gamma_{t-1}} & w_{\beta_{j-1}\beta_j} \\ w_{\beta_j\gamma_1} & \cdots & w_{\beta_j\gamma_{t-1}} & w_{\beta_j\beta_j} \\ w_{\beta_{j+1}\gamma_1} & \cdots & w_{\beta_{j+1}\gamma_{t-1}} & w_{\beta_{j+1}\beta_j} \\ \cdot & \cdots & \cdot & \cdot \\ w_{\beta_t\gamma_1} & \cdots & w_{\beta_t\gamma_{t-1}} & w_{\beta_t\beta_j} \end{vmatrix}.$$

Expanding this out down the final column and using the fact that $w_{\beta_k\beta_l} = \delta_{kl}$, we get

$$v_{\alpha_1,\ldots,\alpha_{r-1},\beta_j,\alpha_{r+1},\ldots,\alpha_m} = (-1)^{t+j} \begin{vmatrix} w_{\beta_1\gamma_1} & \cdots & w_{\beta_1\gamma_{t-1}} \\ \cdot & \cdots & \cdot \\ w_{\beta_{j-1}\gamma_1} & \cdots & w_{\beta_{j-1}\gamma_{t-1}} \\ w_{\beta_{j+1}\gamma_1} & \cdots & w_{\beta_{j+1}\gamma_{t-1}} \\ \cdot & \cdots & \cdot \\ w_{\beta_t\gamma_1} & \cdots & w_{\beta_t\gamma_{t-1}} \end{vmatrix}.$$

Now $v_{1,\ldots,m} = 1$, and we have $i = \beta_j$, $\alpha_r = \gamma_t$ by definition, so that, using the fact that $v_{1,\ldots,i-1,\gamma_t,i+1,\ldots,m} = w_{\beta_j\gamma_t}$, the Plücker relation (8.3) becomes

$$v_{\alpha_1,\ldots,\alpha_m} = \sum_{j=1}^{t} (-1)^{t+j} w_{\beta_j\gamma_t} \begin{vmatrix} w_{\beta_1\gamma_1} & \cdots & w_{\beta_1\gamma_{t-1}} \\ \cdot & \cdots & \cdot \\ w_{\beta_{j-1}\gamma_1} & \cdots & w_{\beta_{j-1}\gamma_{t-1}} \\ w_{\beta_{j+1}\gamma_1} & \cdots & w_{\beta_{j+1}\gamma_{t-1}} \\ w_{\beta_t\gamma_1} & \cdots & w_{\beta_t\gamma_{t-1}} \end{vmatrix}.$$

This can all be put together as

$$v_{\alpha_1,\ldots,\alpha_m} = \begin{vmatrix} w_{\beta_1\gamma_1} & \cdots & w_{\beta_1\gamma_{t-1}} & w_{\beta_1\gamma_t} \\ \cdot & \cdots & \cdot & \cdot \\ w_{\beta_t\gamma_1} & \cdots & w_{\beta_t\gamma_{t-1}} & w_{\beta_t\gamma_t} \end{vmatrix},$$

which proves the assertion. Q.E.D.

It can also be proved that the quadratic relations provide all the equations defining the Grassmann variety; in other words, all the higher degree relations can be expressed in terms of the quadratic Plücker relations.

Exercises to Chapter 8

8.1. Construct a one-to-one correspondence between the 2 dimensional sphere minus one point and the plane.

8.2. Let W be a m dimensional subspace of V and M a frame of W. Prove that for $h \in \mathrm{GL}(m,\mathbb{C})$ the Plücker coordinates determined from hM are equal to $\det M$ times those determined from M.

8.3. Prove that different points of $\mathrm{Grass}(m,V)$ have different Plücker coordinates.

8.4. Prove that the Plücker coordinates do not satisfy any linear relations.

8.5. Use the Laplace expansion to give an alternative proof of the Plücker relations.

9
Infinite dimensional Grassmannians

Chapter 6 showed how the space of all tau functions of the KP hierarchy in Fock space is the orbit of the vacuum state under a group action. In this chapter we show that this orbit is really a Grassmannian, and we consider further the equations which describe it, the bilinear identity. On the way, we touch on the Clifford group and character polynomials.

9.1 The case of finite dimensional Fock space

In the preceding chapter we defined finite dimensional Grassmannians, and explained their Plücker coordinates and the Plücker relations holding between them. This chapter treats the corresponding theory in the framework of Fock spaces, that is, we consider the question of describing the variety of vector subspaces of a fixed dimension in an infinite dimensional vector space. To relate to the material of the preceding chapter, and to avoid technical complications involved in treating the infinite dimensional case, we start by considering the finite dimensional Clifford algebra and the corresponding finite dimensional Fermionic Fock space.

Fix some positive integer N. In this section, from now on, we consider Fermions with indices of absolute value $|i| < N$. Suppose that ψ_i, ψ_i^* are Fermions satisfying the canonical anticommutation relation (4.6), and let $\mathcal{A}_N$ be the finite dimensional Clifford algebra they generate. Write V_N and V_N^* for the vector spaces based by the ψ_i and ψ_i^*, that is,

$$V_N = \bigoplus_{|i|<N} \mathbb{C}\psi_i, \quad V_N^* = \bigoplus_{|i|<N} \mathbb{C}\psi_i^*.$$

Moreover, set

$$W_N = V_N \oplus V_N^*.$$

9.1 The case of finite dimensional Fock space

Let $\mathcal{F}_N$ be the finite dimensional Fermionic Fock space defined in the same way as in Chapter 4. The vacuum state, charge and energy are defined in the same way.

If g is an invertible element of $\mathcal{A}_N$, we define T_g by

$$T_g(a) = gag^{-1} \quad \text{for } a \in \mathcal{A}_N.$$

(Note that $\mathcal{A}_N$ certainly contains noninvertible elements, for example, elements such as $\psi_{1/2}$.) Clearly T_g is an automorphism of $\mathcal{A}_N$, and the following hold:

$$T_{gg'} = T_g T'_g, \quad T_{g^{-1}} = T_g^{-1} \quad \text{and} \quad T_c = 1 \quad \text{for } c \in \mathbb{C} \setminus \{0\}. \quad (9.1)$$

Because of this, the set

$$G(W_N) = \left\{ g \in \mathcal{A}_N \;\middle|\; \begin{array}{l} \text{the inverse } g^{-1} \text{ exists and} \\ T_g(w) \in W_N \text{ for all } w \in W_N \end{array} \right\}$$

forms a group, called the *Clifford group*.

The map taking $w, w' \in W_N$ into the anticommutator $\langle w, w' \rangle = [w, w']_+ \in \mathbb{C}$ is a nondegenerate symmetric bilinear form on W_N. Write

$$O(W_N) = \{ T \in \mathrm{GL}(W_N) \mid \langle T(w), T(w') \rangle = \langle w, w' \rangle \text{ for all } w, w' \in W_N \}$$

for the orthogonal group corresponding to $\langle -, - \rangle$. By definition of the Clifford group, $g \in G(W_N)$ gives $T_g \in O(W_N)$. In fact the following holds.

Theorem 9.1

(i) Any $T \in O(W_N)$ is of the form $T = T_g$ for some $g \in G(W_N)$.
(ii) $T_g = T_{g'}$ if and only if $g = cg'$ with $c \in \mathbb{C} \setminus \{0\}$.

Assertion (i) can be proved using the fact that $O(W_N)$ is generated by reflections; we omit the details. By (9.1), to prove (ii), we need only consider the case $g' = 1$. Now $T_g = 1$ just means that g commutes with every element of $\mathcal{A}_N$, that is, g is in the centre of $\mathcal{A}_n$. However, it is known that the centre of $\mathcal{A}_N$ is $\mathbb{C}$. (See Exercise 9.1.)

We give without proof a formula which allows us to recover g up to a scalar multiples if $T_g \in O(W_N)$ is given. (See, for example, Sato *et al.* (1978, 1979, 1980).) For ease of notation, we set

$$w_i = \begin{cases} \psi_{i-1/2} & \text{for } i = 1, \ldots, N, \\ \psi^*_{i-N-1/2} & \text{for } i = N+1, \ldots, 2N, \\ \psi^*_{2N+1/2-i} & \text{for } i = 2N+1, \ldots, 3N, \\ \psi_{3N+1/2-i} & \text{for } i = 3N+1, \ldots, 4N. \end{cases}$$

We prepare some more notation, setting
$$E_- = \begin{pmatrix} I_{2N} & 0 \\ 0 & 0 \end{pmatrix}, \quad E_+ = \begin{pmatrix} 0 & 0 \\ 0 & I_{2N} \end{pmatrix}, \quad J = \begin{pmatrix} 0 & I_{2N} \\ I_{2N} & 0 \end{pmatrix},$$
where I_{2N} is the $2N \times 2N$ identity matrix.

Now suppose that $T \in O(W_N)$ is given, and assume that $E_+ + E_-T$ is nondegenerate. Then set
$$R = (R_{ij}) = (T-1)(E_+ + E_-T)^{-1}J.$$

In this notation, we find that
$$g = \langle g \rangle : e^\rho : , \quad \text{where} \quad \rho = \frac{1}{2}\sum_{i,j=1}^{4N} R_{ij} w_i w_j,$$
is the required element; that is, $T = T_g$. Here $\langle g \rangle^2 = g^*g \det(E_+ + E_-T)$, where * is Clifford conjugation, the anti-involution of $\mathcal{A}_N$ defined by $w^* = -w$ for $w \in W$. (To say that * is an anti-involution means that $(ab)^* = b^*a^*$ for all $a, b \in \mathcal{A}_N$, so that taking the conjugation * interchanges the order of products.) For $g \in G(W)$ we have $g^*g = gg^* \in \mathbb{C}$. It is possible also to write formulas when $E_+ + E_-T$ is degenerate, but we omit this.

We define the subgroup $\mathbf{G}_N$ of the Clifford group by the following formula:
$$\mathbf{G}_N = \left\{ a \in \mathcal{A} \;\middle|\; \begin{array}{l} \text{there exists } a^{-1} \text{ such that} \\ aV_N a^{-1} = V_N \text{ and } aV_N^* a^{-1} = V_N^* \end{array} \right\}. \tag{9.2}$$

Now suppose given two nondegenerate linear transformations $V_N \to V_N$ and $V_N^* \to V_N^*$ that satisfy
$$\psi_i \mapsto \sum_j a_{ji}\psi_j \quad \text{and} \quad \psi_{-i}^* \mapsto \sum_j b_{ij}\psi_{-j}, \quad \text{with} \quad (b_{ij}) = (a_{ij})^{-1}. \tag{9.3}$$

If we recall the definition of the bilinear form $\langle -, - \rangle$ on W_N, we see that these two linear transformations together define an orthogonal transformation of W_N.

Now given an element $|u\rangle \in \mathcal{F}_N$, we set
$$V_N(|u\rangle) = \{v \in V_N \mid v|u\rangle = 0\}.$$
This is a vector subspace of V_N. For example, we have
$$V_N(|\text{vac}\rangle) = \bigoplus_{0 < i < N} \mathbb{C}\psi_i;$$

9.1 The case of finite dimensional Fock space

more generally if $|u\rangle = \psi_{m_1}\cdots\psi_{m_r}\psi^*_{n_1}\cdots\psi^*_{n_r}|\text{vac}\rangle$ with $m_1 < \cdots < m_r < 0$ and $n_1 < \cdots < n_r < 0$ then

$$V_N(|u\rangle) = \left(\bigoplus_{\substack{0<i<N \\ i\neq -n_1,\ldots,-n_r}} \mathbb{C}\psi_i\right) \oplus \mathbb{C}\psi_{m_1} \oplus \cdots \oplus \mathbb{C}\psi_{m_r}.$$

This is N dimensional, that is, half the dimension of V_N.

However, for example, if $|u\rangle = (\psi_{-1/2}\psi^*_{-1/2} + \psi_{-3/2}\psi^*_{-3/2})|\text{vac}\rangle$ then we have

$$V_N(|u\rangle) = \bigoplus_{5/2 \leq i \leq N} \mathbb{C}\psi_i.$$

This has dimension $N-2$, so its dimension has dropped from that in the previous examples. Quite generally, if $|u\rangle$ has charge 0 then $\dim V_N(|u\rangle) \leq N$. (See Exercise 9.2.)

In what follows, we consider the case $|u\rangle = g|\text{vac}\rangle$ for $g \in \mathbf{G}_N$.

Lemma 9.2 *For any $g, g' \in \mathbf{G}_N$ we have*

$$V_N(g|\text{vac}\rangle) = V_N(g'|\text{vac}\rangle) \iff g = cg' \quad \text{for some } c \in \mathbb{C} \setminus \{0\}. \tag{9.4}$$

Proof Suppose that $g \in \mathbf{G}_N$. For $u \in V_N(g|\text{vac}\rangle)$ we have $ug|\text{vac}\rangle = 0$. Moreover, observing that $ug = gT_{g^{-1}}(u)$, we see that

$$V_N(g|\text{vac}\rangle) = T_g(V_N(|\text{vac}\rangle)). \tag{9.5}$$

(In fact this identity holds for any $|v\rangle \in \mathcal{F}_N$ in place of $|\text{vac}\rangle$.) Therefore for the lemma, it is enough to prove (9.4) in the case $g' = 1$. Suppose that $g \in \mathbf{G}_N$ and that

$$V_N(g|\text{vac}\rangle) = V_N(|\text{vac}\rangle).$$

Then $\psi_i g|\text{vac}\rangle = 0$ for $i > 0$. As we have already mentioned, $\mathbf{G}_N \cong \mathrm{GL}(V_N)$, so that using (9.2) and the following relation (9.3), we see that also $\psi^*_i g|\text{vac}\rangle = 0$ for $i > 0$. From the properties of the vacuum state it follows from this that $g|\text{vac}\rangle$ is a scalar multiple of $|\text{vac}\rangle$. Q.E.D.

Now the action of the group $\mathrm{GL}(V_N)$ of nondegenerate linear transformations of V_N takes $V_N(|\text{vac}\rangle)$ to a vector subspace of the same dimension N, and conversely, each subspace of V_N of the same dimension is the image of $V_N(|\text{vac}\rangle)$ under an element of $\mathrm{GL}(V_N)$. Thus we see that the orbit $\mathbf{G}_N|\text{vac}\rangle$ of the vacuum state can be identified with the

Grassmannian of N dimensional vector subspaces of V_N. The equations defining the Grassmannian are the Plücker relations, so that for the Grasmanian discussed above, the Plücker relations are at the same time the equations defining the orbit of the vacuum state. In the following section we give the equation describing the orbit of the vacuum vector in the infinite dimensional case, which includes the case of finite dimensional Fock space.

9.2 Description of the vacuum orbit

At the risk of repeating ourselves, we take up again the bilinear identity of Chapter 6, adding explanations of some of the material left over from there.

Theorem 9.3 (Bilinear identity) *An element $|u\rangle$ of Fock space having charge 0 belongs to the orbit of the vacuum state if and only if*

$$\sum_{i \in \mathbb{Z}+1/2} \psi_i^* |u\rangle \otimes \psi_{-i} |u\rangle = 0. \tag{9.6}$$

The first half of the proof we now give is the same as the proof in Chapter 6, but we repeat it for convenience.

Proof In the case $|u\rangle = |\text{vac}\rangle$, the ψ_i^* and ψ_{-i} are annihilation operators, so (9.6) holds obviously.

Let X_A be an element of the Lie algebra $\mathfrak{gl}(\infty)$ having the following commutation relations with the Fermions (see (6.12)):

$$[X_A, \psi_i^*] = \sum_j (-a_{ij}) \psi_j^* \quad \text{and} \quad [X_A, \psi_{-i}] = \sum_j (a_{ji}) \psi_{-j}.$$

From the above relation we get

$$\sum_{i \in \mathbb{Z}+1/2} [X_A, \psi_i^*] |\text{vac}\rangle \otimes \psi_{-i} |\text{vac}\rangle + \sum_{i \in \mathbb{Z}+1/2} \psi_i^* |\text{vac}\rangle \otimes [X_A, \psi_{-i}] |\text{vac}\rangle = 0.$$

It follows from this that the relation (9.6) is invariant under the action of e^{X_A}. Thus (9.6) holds if $|u\rangle$ is a point of the orbit of the vacuum vector.

We now prove the converse. We start by noting the following identity: if we set $\phi^* = \psi_{-i}^* + \psi_j^*$ and $\phi = \psi_i + \psi_{-j}$ then

$$[\phi, \phi^*]_+ = 2 \quad \text{and} \quad e^{\pi i \phi^* \phi / 2} = 1 - \phi^* \phi \in \mathbf{G}.$$

9.2 Description of the vacuum orbit

Using this identity, we get

$$\left(1 - (\psi^*_{-m_1} + \psi^*_{n_1})(\psi_{m_1} + \psi_{-n_1})\right)\psi_{m_1}\cdots\psi_{m_r}\psi^*_{n_1}\cdots\psi^*_{n_r}|\text{vac}\rangle$$
$$= (-1)^{r-1}\psi_{m_2}\cdots\psi_{m_r}\psi^*_{n_2}\cdots\psi^*_{n_r}|\text{vac}\rangle, \qquad (9.7)$$

say, for any $m_1 < \cdots < m_r < 0$ and $n_1 < \cdots < n_r < 0$.

Now let $|u\rangle$ be any element of $\mathcal{F}$ of charge 0. Then $|u\rangle$ is a linear combination of terms, each of which consists of the vacuum state $|\text{vac}\rangle$ acted on by the same number of Fermions ψ_i and ψ^*_j. Among these terms, we pick out one with the smallest number of ψ_i; if there are several such terms, we choose one arbitrarily (for example, we could order these terms lexicographically by the increasing indices of the ψ, and choose the smallest). This term involves the same number of ψ_i and ψ^*_j, so that we can successively reduce the number of pairs using the relation (9.7). Thus any vector $|u\rangle \in (\mathcal{F}_N)_0$ can be transformed by a suitable element of the group $\mathbf{G}$ to an element of the form

$$|\text{vac}\rangle + \sum_{i,j<0} c_{ij}\psi_i\psi^*_j|\text{vac}\rangle + \cdots,$$

where $\cdots$ stands for a sum of terms involving multiplication by at least four Fermions. Next, acting by an element of $\mathbf{G}$ of the form $\exp(-\sum_{i,j<0} c_{ij}\psi_i\psi^*_j)$, we can reduce it to the form

$$|u'\rangle = |\text{vac}\rangle + \sum_{i,j,k,l<0} c_{ijkl}\psi_i\psi_j\psi^*_k\psi^*_l|\text{vac}\rangle + \cdots.$$

As we saw above, the relation (9.6) is invariant under the action of $\mathbf{G}$, so that we have

$$\sum_{i\in\mathbb{Z}} \psi^*_i|u'\rangle \otimes \psi_{-i}|u'\rangle = 0.$$

From the form of $|u'\rangle$ we clearly have $\psi^*_i|u'\rangle \neq 0$ for $i < 0$, so that

$$\psi_{-i}|u'\rangle = 0 \quad \text{for } -i > 0,$$

and similarly

$$\psi^*_i|u'\rangle = 0 \quad \text{for } i > 0.$$

This proves that $|u'\rangle$ is a scalar multiple of $|\text{vac}\rangle$. Q.E.D.

Putting this together with the material of the previous section, we see that the bilinear identity (9.6) is basically the same thing as the Plücker relations. In the following chapter, Chapter 10, we rewrite this relation via the Boson–Fermion correspondence to obtain another derivation of

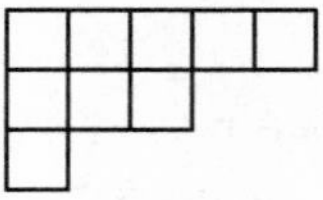

Fig. 9.1. Young diagram I. $Y = (5, 3, 1)$

the Plücker relations of Chapter 8. As a by-product, this gives rise to connections with other areas of mathematics. For this we need some minor preliminaries, which we treat in the following section.

9.3 Young diagrams and character polynomials

We now make a small digression, introducing some preliminary material which in the following Chapter 10 will help us to understand the connection between the relation of the preceding section and the Plücker relations. We introduce the *character polynomials*, which form one possible basis for the polynomial ring. Pursuing this discussion further in one direction leads to another area of mathematics, namely, combinatorics. The aim of this section is to explain how under the Boson–Fermion correspondence, the basis (4.11) of Fock space $(\mathcal{F}_N)_0$ of charge 0 goes over into the character polynomials.

We start by explaining the character polynomials. The objects we describe are in essence the same thing as the Maya diagrams; however, for our purpose, we introduce the notion of *Young diagrams*, which appeared earlier in the mathematics literature. Young diagrams can be expressed in a number of different ways; one is to say that a Young diagram is a nonincreasing sequence of positive integers $(f_1, \ldots, f_r)$. In pictorial form, a Young diagram is viewed as a figure in the fourth quadrant of the plane, made up of a number of rows of congruent square tiles, with the rows aligned along their left sides, the first row having f_1 tiles, the second row f_2 tiles, and so on. The only requirement is that the number of tiles in a row does not increase when we move down from one row to the next (see Figure 9.1).

It is known that if we limit the number of rows, say to diagrams with at most n rows, then Young diagrams parametrise the irreducible representations of the general linear group $GL(n, \mathbb{C})$. (See for example W. Fulton and J. Harris, *Representation theory*, Chap. 15.) The Young diagrams also relate to many interesting problems of combinatorics.

9.3 Young diagrams and character polynomials

Fig. 9.2. Young diagram II. $Y = (4,1|2,0)$

Young diagrams have the following alternative description (see Figure 9.2).

Suppose that $Y = (f_1, \ldots, f_r)$ is a Young diagram, and let s be the diagonal width of Y when viewed from the top left-hand corner. We write $m_1 > \cdots > m_s$ for the number of tiles lying above the NW–SE diagonal line (excluding those straddling the line) in each horizontal row, and $n_1 > \cdots > n_s$ for the number of tiles lying below the diagonal line (excluding those straddling the line) in each vertical column. Then, we write

$$Y = (m_1, \ldots, m_s \mid n_1, \ldots, n_s)$$

for the Young diagram.

Using this notation, we define the *character polynomial* of Y to be

$$\chi_Y(\mathbf{x}) = \det(h_{m_i n_j}(\mathbf{x})).$$

Here

$$h_{mn}(\mathbf{x}) = (-1)^n \sum_{l \geq 0} p_{l+m+1}(\mathbf{x}) p_{n-l}(-\mathbf{x}) \tag{9.8}$$

$$= (-1)^{n+1} \sum_{l < 0} p_{l+m+1}(\mathbf{x}) p_{n-l}(-\mathbf{x}), \tag{9.9}$$

where the $p_i(\mathbf{x})$ are defined in (5.20). (We set $p_i(\mathbf{x}) = 0$ if $i < 0$.) It is convenient to set also

$$q_n(\mathbf{x}) = (-1)^n p_n(-\mathbf{x}).$$

Here $h_{mn}(\mathbf{x}) = \chi_{(m+1,1^n)}(\mathbf{x})$ is the character polynomial corresponding to the hook shaped Young diagram $(m+1, 1^n)$, where 1^n stands for the series of n terms $(1, 1, \ldots, 1)$. While we are on the subject, $p_n(\mathbf{x}) = \chi_{(n)}(\mathbf{x})$ is the character polynomial corresponding to a Young diagram consisting of a single row of n tiles, and $q_n(\mathbf{x}) = \chi_{(1^n)}(\mathbf{x})$ is that corresponding to the Young diagram (1^n) consisting of a column of n single tiles.

Example 9.1 *We give a few examples, distinguishing the Young diagram by writing it as a subscript on* χ:

$$\chi_\emptyset(\mathbf{x}) = 1,$$
$$\chi_{(1)}(\mathbf{x}) = x_1,$$
$$\chi_{(2)}(\mathbf{x}) = \frac{x_1^2}{2} + x_2,$$
$$\chi_{(1,1)}(\mathbf{x}) = \frac{x_1^2}{2} - x_2,$$
$$\chi_{(2,1)}(\mathbf{x}) = \frac{x_1^3}{3} - x_3,$$
$$\chi_{(2,2)}(\mathbf{x}) = \frac{x_1^4}{12} - x_1 x_3 + x_2^2.$$

For the sake of brevity, we do not go into detailed explanations at this point, leaving proofs to specialist textbooks (see for example I.G. Macdonald, *Symmetric functions and Hall polynomials*, Oxford, 1979, second edition 1995); however, the character polynomial of a Young diagram Y is the function obtained from the representation matrix of the irreducible character of the general linear group $GL(n, \mathbb{C})$ determined by Y by writing out the sum of powers of the eigenvalues as a function of the elementary symmetric functions. (This is independendent of the size n of the matrices, provided that it is bigger than the number of rows of the Young diagram.) In what follows, the $\chi_Y(\mathbf{x})$ are called character polynomials. It is known that the character polynomials form a basis of the polynomial ring.

The following theorem plays an important part in the arguments of the following chapter.

Theorem 9.4 *Under the Boson–Fermion correspondence, the basis vector*

$$\psi_{m_1} \cdots \psi_{m_r} \psi^*_{n_1} \cdots \psi^*_{n_r} |\mathrm{vac}\rangle \text{ for } m_1 < \cdots < m_r < 0 \text{ and } n_1 < \cdots < n_r < 0$$

of the Fock space of charge 0 goes over into the character polynomial of the Young diagram of the form

$$Y = (-m_1 - 1/2, \ldots, -m_r - 1/2 \mid -n_1 - 1/2, \ldots, -n_r - 1/2)$$

multiplied by the sign $(-1)^{\sum\limits_{i=1}^{r}(n_i+1/2)+r(r-1)/2}$.

For the proof, we prepare the lemma below: write the time evolution

9.3 Young diagrams and character polynomials

of the Fermions as

$$\psi_n(\mathbf{x}) = e^{H(\mathbf{x})}\psi_n e^{-H(\mathbf{x})}, \quad \psi_n^*(\mathbf{x}) = e^{H(\mathbf{x})}\psi_n^* e^{-H(\mathbf{x})}.$$

By the formulas (5.21) and (5.22) of Chapter 5, we can write

$$\psi_n(\mathbf{x}) = \sum_{j=0}^{\infty}\psi_{n+j}p_j(\mathbf{x}) \quad \text{and} \quad \psi_n^*(\mathbf{x}) = \sum_{j=0}^{\infty}\psi_{n+j}^*p_j(-\mathbf{x}).$$

Lemma 9.5 *For $m, n > 0$ we have*

$$h_{mn}(\mathbf{x}) = (-1)^n \langle \mathrm{vac}|\psi_{-m-1/2}(\mathbf{x})\psi_{-n-1/2}^*(\mathbf{x})|\mathrm{vac}\rangle.$$

Proof Substituting the above formulas gives

$$\begin{aligned}
\langle \mathrm{vac}|\psi_i(\mathbf{x})\psi_j^*(\mathbf{x})|\mathrm{vac}\rangle &= \sum_{s,t=0}^{\infty} \langle \mathrm{vac}|\psi_{i+s}\psi_{j+t}^*|\mathrm{vac}\rangle p_s(\mathbf{x})p_t(-\mathbf{x}) \\
&= \sum_{-s<i} \sum_{0\leq t<-j} p_s(\mathbf{x})p_t(-\mathbf{x})\delta_{i+j+s+t,0} \\
&= \sum_{t=0}^{-j-1/2} p_{-i-j-t}(\mathbf{x})p_t(-\mathbf{x}) \\
&= \sum_{l=0}^{-j-1/2} p_{-i+1/2+l}(\mathbf{x})p_{-j-1/2-l}(-\mathbf{x}) \\
&= (-1)^{j+1/2} h_{-i-1/2,-j-1/2}(\mathbf{x}). \quad \text{Q.E.D.}
\end{aligned}$$

For the proof of Theorem 9.4, we only need to compare the definition of the character polynomial and Wick's theorem.

The Young diagrams appearing here can be expressed more simply in terms of Maya diagrams. We now explain the relation between Maya diagrams and Young diagrams (see Figure 9.3). We hope that the correspondence in the general case should be clear from the following example. Consider the Maya diagram $\mathbf{m} = \{-7/2, -5/2, -1/2, 3/2, 7/2, \ldots\}$, where ... means that from that point on, every half-integer is black in Figure 9.3.

We now imagine the line as a ruler with ticks marked off at every integer point, and a white or black stone at every half-integer midway between the ticks. We fold up the line depending on the white and black stones according to the following rule: we start from the $i \ll 0$ part of the Maya diagram. By assumption, the stones at all points $i < -7/2$ are white. For this white segment, we draw a straight line vertically upwards

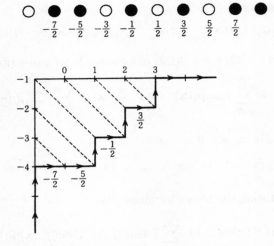

Fig. 9.3. The Maya diagram $\mathbf{m} = \{-7/2, -5/2, -1/2, 3/2, 7/2, \ldots\}$ and its Young diagram

starting from $-\infty$, and heading upwards one unit for each white stone. For the black segment, we draw a straight line from left to right, heading one unit to the right for each black stone out to $+\infty$. In the present example, we proceed straight up until the position -4. Corresponding to the two black stones at $-7/2$ and $-5/2$, we head two steps across to the right from -4 to -2. Then one step up (for $-3/2$), one step right (for $-1/2$), up (for $1/2$), right (for $3/2$), up (for $5/2$). From $7/2$ onwards all the stones are black, so that from 3 onwards we proceed horizontally right.

In this way, any Young diagram can be determined as the region of the plane bounded by two line segments and a bent line. The charge of a Maya diagram can be expressed as being $i + 1/2$, where the two half-lines meet at the midpoint of the interval from i to $i + 1$. In the present case, the charge is -1 (see Figure 9.3). A Young diagram in the usual sense can be treated as a charged Young diagram by viewing it in the fourth quadrant of the plane, taking the top left corner as the origin, and then translating up or down (not sideways) to achieve the desired charge. To pass from a charged Young diagram to a Maya diagram, we just retrace our steps in the procedure described above. In this way we see that there is a one-to-one correspondence between charged Young diagrams and Maya diagrams.

In general, consider a Maya diagram with white stones at $-m_r < \cdots < -m_1$ among positions with $i > 0$, and black stones at $n_1 < \cdots < n_s$ among positions with $i < 0$. Then the corresponding Young diagram is described as follows. The diagonal width is $\max(r, s)$. If $r \geq s$, the number of tiles to the right of the diagonal line in each successive row is

$$-m_1 - 1/2 - (r - s) > \cdots > -m_r - 1/2 - (r - s),$$

and the number of tiles below the diagonal line in each column is

$$-n_1 - 1/2 + (r - s) > \cdots > -n_s - 1/2 + (r - s) > r - s - 1 > \cdots > 0.$$

Similarly, if $s \geq r$, the number of tiles to the right of the diagonal line in each successive row is

$$-m_1 + 1/2 + (s - r) > \cdots > -m_r + 1/2 + (s - r) > s - r > \cdots > 1,$$

and the number of tiles in each column below the diagonal line is

$$-n_1 - 1/2 - (s - r) > \cdots > -n_s - 1/2 - (s - r).$$

Exercises to Chapter 9

9.1. Prove that the finite dimensional Clifford algebra $\mathcal{A}(W_N)$ introduced in Section 9.1 is isomorphic to the algebra of all $2^{2N} \times 2^{2N}$ matrices (compare Chapter 4, Exercise 4.1). Using this, deduce that the centre of $\mathcal{A}(W_N)$ is $\mathbb{C}$.

9.2. In a finite dimensional Fock space, prove that $\dim V_N(|u\rangle) \leq N$ for an element $|u\rangle$ of charge 0.

9.3. Verify the relation between Young diagrams and Maya diagrams explained in Section 9.3 in the particular case

$$\mathbf{m} = \{-11/2, -7/2, -3/2, -1/2, 5/2, 7/2, 13/2, \ldots\}.$$

9.4. Determine the character polynomials corresponding to the following Young diagrams:

   (a) $Y_1 = (3, 1)$;
   (b) $Y_1 = (3, 2, 1)$;
   (c) $Y_1 = (4, 2, 1)$.

10
The bilinear identity revisited

In this chapter, we show how the bilinear identity discussed in previous chapters can be rewritten as the Plücker relations for an infinite dimensional Grassmannian. The chapter can also be seen as an exercise in applying Wick's theorem. Moreover, we derive the Hirota equation from the Plücker relations.

10.1 The bilinear identity and the Plücker relations

Let $|u\rangle \in \mathcal{F}_0$ be an element of the Fock space of charge 0, and

$$f(\mathbf{x}; |u\rangle) = \langle \text{vac}|e^{H(\mathbf{x})}|u\rangle$$

the element of the Bosonic Fock space corresponding to $|u\rangle$ under the Boson–Fermion correspondence. Because the character polynomials $\chi_Y(\mathbf{x})$ form a basis of the polynomial ring, we can write $f(\mathbf{x}; |u\rangle)$ as a linear combination of these:

$$f(\mathbf{x}; |u\rangle) = \sum_Y c_Y(|u\rangle)\chi_Y(\mathbf{x}).$$

Here $\chi_Y(\mathbf{x})$ is the character polynomial corresponding to the Young diagram Y, and the sum runs over the set of all Young diagrams. In this section, we show that the necessary and sufficient condition for the left-hand side to be a tau function of the KP hierarchy is that the coefficients $c_Y(|u\rangle)$ satisfy the Plücker relations.

As we saw in Chapter 4, the Fock space $\mathcal{F}_0$ of charge 0 is spanned by vectors of the form

$$f(\mathbf{m}, \mathbf{n}) = \psi_{m_1} \cdots \psi_{m_r} \psi^*_{n_1} \cdots \psi^*_{n_r}|\text{vac}\rangle,$$

where we set $\mathbf{m} = (m_1, \ldots, m_r)$ and $\mathbf{n} = (n_1, \ldots, n_r)$. However, for our

10.1 The bilinear identity and the Plücker relations

purposes in this chapter, this parametrisation of the basis vectors is not appropriate, because it leads to many irrelevant signs in the formulas. In what follows, we index the basis vectors in the Fock space $\mathcal{F}_l$ by Maya diagrams $\alpha = (\alpha_j)_{j \geq 1}$. In fact, we slightly extend the notion of Maya diagrams, as follows. In Chapter 4, a sequence of half-integers α was called a Maya diagram if

(i) $\alpha_j < \alpha_{j+1}$ for all $j \geq 1$, and
(ii) $\alpha_j + 1 = \alpha_{j+1}$ for sufficiently large j.

Here, we ignore (i), and require only condition (ii); we call α a *signed Maya diagram*.

Let α be a signed Maya diagram. Set

$$l = \lim_{j \to \infty} \left(j - \alpha_j - \frac{1}{2} \right).$$

The integer l is called the *charge* of α. We associate a vector $|\alpha\rangle \in \mathcal{F}_l$ with α. Specifically, $|\alpha\rangle$ is a vector in the Fock space of charge l.
If (i) holds, we set

$$|\alpha\rangle = \psi^*_{\alpha_1} \psi^*_{\alpha_2} \psi^*_{\alpha_3} \cdots |\Omega\rangle. \tag{10.1}$$

Suppose that α is written as

$$\alpha = \left(n_1, \ldots, n_r, \frac{1}{2}, \ldots, -m_r - 1, -m_r + 1, \ldots, -m_1 - 1, m_1 + 1, \ldots \right).$$

Then $|\alpha\rangle$ equals $\psi_{m_1} \cdots \psi_{m_r} \psi^*_{n_1} \cdots \psi^*_{n_r} |\text{vac}\rangle$ up to sign.

We set $|\alpha\rangle = 0$ if $\alpha_j = \alpha_k$ for some $j \neq k$. Otherwise, by setting $\beta_j = \alpha_{\sigma(j)}$ where $\sigma : \{1, 2, \ldots\} \to \{1, 2, \ldots\}$ is an appropriate permutation satisfying $\sigma(j) = j$ for almost all j, we obtain a Maya diagram β satisfying condition (i). We then set $|\alpha\rangle = \text{sign}(\sigma)|\beta\rangle$.

In what follows, for brevity we say that α is a Maya diagram even if it does not satisfy condition (i). For a half-integer b and a Maya diagram α, we denote by $\alpha \oplus b$ the Maya diagram $(b, \alpha_1, \alpha_2, \ldots)$. For a Maya diagram α, we denote by $\alpha \ominus \alpha_j$ the Maya diagram $(\alpha_1, \ldots, \alpha_{j-1}, \alpha_{j+1}, \ldots)$. If $|\alpha\rangle \in \mathcal{F}_l$, then $|\alpha \oplus b\rangle \in \mathcal{F}_{l+1}$ and $|\alpha \ominus \alpha_j\rangle \in \mathcal{F}_{l-1}$.

As we discussed in Section 9.2, Theorem 9.3, the necessary and sufficient condition for

$$f = \sum_\alpha c(\alpha)|\alpha\rangle$$

to be in the group orbit of the vacuum state |vac⟩ is that

$$\sum_{i\in\mathbb{Z}+1/2} \psi_i^* f \otimes \psi_{-i} f = 0. \tag{10.2}$$

Here, we restrict the summation to the Maya diagrams in the strict sense, and of charge 0. It is convenient to extend the notation $c(\alpha)$ to an arbitrary signed Maya diagram. We set $c(\alpha) = 0$ if $|\alpha\rangle = 0$. For α and β as above, we set

$$c(\alpha) = \text{sign}(\sigma)c(\beta).$$

We now state the main result of this chapter. We rephrase condition (10.2) in terms of the coefficients $c(\alpha)$. Let γ be a Maya diagram of charge 1 and δ of charge -1. Then, condition (10.2) is equivalent to

$$\sum_{j=1}^{\infty} (-1)^j c(\gamma \ominus \gamma_j) c(\delta \oplus \gamma_j) = 0. \tag{10.3}$$

The proof is straightforward, and we leave it to the reader. We write out an example: for $\gamma = \{-1/2, 1/2, 3/2, \ldots\}$ and $\delta = \{-3/2, 5/2, 7/2, \ldots\}$, we get

$$\begin{aligned}
0 = {}& c(\{1/2, 3/2, 5/2, \ldots\}) \, c(\{-3/2, -1/2, 5/2, 7/2, \ldots\}) \\
& -c(\{-1/2, 3/2, 5/2, \ldots\}) \, c(\{-3/2, 1/2, 5/2, 7/2, \ldots\}) \\
& +c(\{-1/2, 1/2, 5/2, 7/2, \ldots\}) \, c(\{-3/2, 3/2, 5/2, 7/2, \ldots\}).
\end{aligned} \tag{10.4}$$

10.2 Plücker relations and the Hirota equation

Summarising the results of the preceding section, we continue by showing that the Plücker relations obtained here also imply the bilinear differential equations of Hirota form.

For an element $g \in \mathbf{G}$, we let

$$\tau(\mathbf{x}; g) = \langle \text{vac} | e^{H(\mathbf{x})} g | \text{vac} \rangle$$

be the element of the Bosonic Fock space obtained under the Boson–Fermion correspondence. This is a tau function of the KP hierarchy. Because the character polynomials form a basis of the polynomial ring, we can write $\tau(\mathbf{x}; g)$ as a linear combination of these:

$$\tau(\mathbf{x}; g) = \sum_Y c_Y(g) \chi_Y(\mathbf{x}).$$

Here $\chi_Y(\mathbf{x})$ is the character polynomial corresponding to the Young

10.2 Plücker relations and the Hirota equation

diagram Y, and the sum runs over the set of all Young diagrams. The result we proved in the preceding section is that the coefficients $c_Y(g)$ satisfy the Plücker relations. In what follows, we call these coefficients $c_Y(g)$ the *Plücker coordinates* of the tau function $\tau(\mathbf{x};g)$.

Now $\tau(\mathbf{x};g)$ is an element of the polynomial ring $\mathbb{C}[x_1,x_2,\ldots]$. We write ∂_i for the partial derivatives with respect to the x_i, and $\mathbb{C}[\partial_1,\partial_2,\ldots]$ for the polynomial ring generated by the ∂_i; this is the commutative ring of all differential operators with constant coefficients in the independent variable $x_1, x_2, \ldots$. For a polynomial $p(\mathbf{x})$, we write $p(\widetilde{\partial}_\mathbf{x}) \in \mathbb{C}[\partial_1,\partial_2,\ldots]$ for the element obtained by making the substitution $x_i \mapsto (1/i)\partial_i$. The formula

$$\langle p(\widetilde{\partial}_\mathbf{x}), f(\mathbf{x})\rangle = p(\widetilde{\partial}_\mathbf{x})f(\mathbf{x})|_{\mathbf{x}\to 0} \qquad (10.5)$$

defines a nondegenerate pairing between the polynomial rings $\mathbb{C}[\partial_1,\partial_2,\ldots]$ and $\mathbb{C}[x_1,x_2,\ldots]$. One sees easily that under this pairing, the elements

$$\left\{\frac{x_1^{m_1}}{m_1!}\frac{x_2^{m_2}}{m_2!}\cdots\right\}_{m_1\geq 0, m_2\geq 0,\ldots} \quad \text{and} \quad \{\partial_1^{m_1}\partial_2^{m_2}\}_{m_1\geq 0, m_2\geq 0,\ldots}$$

form dual bases. In particular, if we give x_i degree i as in Chapter 4, and moreover give ∂_i degree $-i$, then we have decompositions into homogeneous components

$$\mathbb{C}[x_1,x_2,\ldots] = \bigoplus_{n\geq 0}\mathbb{C}[\mathbf{x}]_n \quad \text{and} \quad \mathbb{C}[\partial_1,\partial_2\ldots] = \bigoplus_{n\geq 0}\mathbb{C}[\partial_\mathbf{x}]_{-n},$$

and the pairing $\langle p(\widetilde{\partial}_\mathbf{x}), f(\mathbf{x})\rangle$ gives a nondegenerate pairing between $\mathbb{C}[\mathbf{x}]_n$ and $\mathbb{C}[\partial_\mathbf{x}]_{-n}$, with $\langle \mathbb{C}[\partial_\mathbf{x}]_{-m}, \mathbb{C}[\mathbf{x}]_n\rangle = 0$ for $m \neq n$.

It is known that if we pass to the basis of $\mathbb{C}[x_1,x_2,\ldots]$ formed by the character polynomials and reinterpret the pairing in these terms, then the following orthogonality relation holds (see Exercise 10.3).

Lemma 10.1 *For any two Young diagrams Y, Y', we have*

$$\langle \chi_Y(\widetilde{\partial}_\mathbf{x}), \chi_{Y'}(\mathbf{x})\rangle = \delta_{YY'}.$$

In other words, any $f(\mathbf{x}) \in \mathbb{C}[x_1,x_2,\ldots]$ can be written as a linear combination of character polynomials

$$f(\mathbf{x}) = \sum_Y c_Y \chi_Y(\mathbf{x})$$

and the coefficients c_Y can be computed using the above pairing:

$$c_Y = \langle \chi_Y(\widetilde{\partial}_\mathbf{x}), f(\mathbf{x})\rangle.$$

We can deduce a number of interesting consequences from this orthogonality relation. For $g \in \mathbf{G}$, set $g(\mathbf{x}) = e^{H(\mathbf{x})} g e^{-H(\mathbf{x})}$, and consider the transformation of the following type of expression:

$$\begin{aligned}\tau(\mathbf{x}+\mathbf{y};g) &= \langle \text{vac}|e^{H(\mathbf{x}+\mathbf{y})}g|\text{vac}\rangle \\ &= \langle \text{vac}|e^{H(\mathbf{x})}e^{H(\mathbf{y})}g|\text{vac}\rangle \\ &= \tau(\mathbf{x};g(\mathbf{y})) \\ &= \sum_Y c_Y(g(\mathbf{y}))\chi_Y(\mathbf{x}).\end{aligned}$$

Then by Lemma 10.1 above, we get

$$\begin{aligned}c_Y(g(\mathbf{y})) &= \langle \chi_Y(\widetilde{\partial}_\mathbf{x}), \tau(\mathbf{x}+\mathbf{y};g)\rangle \\ &= \chi_Y(\widetilde{\partial}_\mathbf{x})\tau(\mathbf{x}+\mathbf{y};g)|_{\mathbf{x}\to 0}.\end{aligned}$$

Using the fact that $\dfrac{\partial \tau(\mathbf{x}+\mathbf{y};g)}{\partial x_i} = \dfrac{\partial \tau(\mathbf{x}+\mathbf{y};g)}{\partial y_i}$, and putting this together with

$$\tau(\mathbf{x}+\mathbf{y};g) = \sum_Y c_Y(g)\chi_Y(\mathbf{x}+\mathbf{y}),$$

we obtain

$$\begin{aligned}c_Y(g(\mathbf{y})) &= \chi_Y(\widetilde{\partial}_\mathbf{y})\tau(\mathbf{x}+\mathbf{y};g)|_{\mathbf{x}\to 0} \\ &= \chi_Y(\widetilde{\partial}_\mathbf{y})\tau(\mathbf{y};g).\end{aligned}$$

This gives the following lemma.

Lemma 10.2 *For an element $g \in \mathbf{G}$, write $c_Y(g)$ for the Plücker coordinates of $\tau(\mathbf{x};g)$, and consider the time evolution $g(\mathbf{x}) = e^{H(\mathbf{x})}ge^{-H(\mathbf{x})}$ of g. Let $\tau(\mathbf{x};g)$ be the tau function of the corresponding KP hierarchy. Then the following relations hold:*

(i) $c_Y(g) = \chi_Y(\widetilde{\partial}_\mathbf{x})\tau(\mathbf{x};g)|_{x\to 0}$, and

(ii) $c_Y(g(\mathbf{x})) = \chi_Y(\widetilde{\partial}_\mathbf{x})\tau(\mathbf{x};g)$.

In particular, $\{c_Y(g(\mathbf{x}))\}$ is again the Plücker coordinate of a tau function of the KP hierarchy.

Combining the fact that the Plücker coordinates satisfy the Plücker relation together with the fact that, as shown by (ii) of the lemma, the Plücker coordinates can be expressed in terms of the tau function, we can also deduce the Hirota bilinear differential equation.

10.2 Plücker relations and the Hirota equation

Example 10.1 *As an example, let us try to write out the Plücker relation from the end of Section 10.1 using the tau function. By the table of character polynomials given in the example of the preceding chapter and by the above lemma, in the present case, the Plücker relation can be rewritten as follows in terms of the tau function as a differential equation:*

$$0 = \frac{1}{12}\tau(\mathbf{x}) \cdot (\partial_1^4 - 4\partial_1\partial_3 + 3\partial_2^2)\tau(\mathbf{x})$$
$$-\frac{1}{3}\partial_1\tau(\mathbf{x}) \cdot (\partial_1^3 - \partial_3)\tau(\mathbf{x})$$
$$+\frac{1}{4}(\partial_1^2 + \partial_2)\tau(\mathbf{x}) \cdot (\partial_1^2 - \partial_2)\tau(\mathbf{x}).$$

If we transform this differential equation into Hirota form, we get the equation

$$(D_1^4 + 3D_2^2 - 4D_1D_3)\tau \cdot \tau = 0.$$

which has already appeared in Chapter 3.

Exercises to Chapter 10

10.1. Prove equation (10.3) of the main text.

10.2. Show that the Plücker relation (10.4) at the end of Section 10.1 is the same thing as the Plücker relation of Section 8.3, and think about why this must be true.

10.3. Find the matrix describing the change of basis between the two bases

$$\{x_3, x_2x_1, x_1^3/6\}, \quad \text{and} \quad \{\chi_{(3)}(\mathbf{x}), \chi_{(2,1)}(\mathbf{x}), \chi_{(1,1,1)}(\mathbf{x})\}$$

of the space of cubic polynomials. Verify (10.2) in this case.

10.4. Determine the Plücker relation that gives rise to the differential equation in Hirota bilinear form

$$(D_1^3D_2 + 2D_2D_3 - 3D_1D_4)\tau \cdot \tau = 0.$$

Solutions to exercises

Chapter 1

1.1. The question is to find $F(x,t)$ such that
$$F(x,0) = x \quad \text{and} \quad \partial F/\partial t = x^2(\partial F/\partial x).$$
Expand in powers of t as $F(x,0) = x + ta_1 + t^2 a_2 + \cdots$ and substitute in the equation to get $a_1 = x^2$, etc. Now guess a closed form. The answer is $x/(1-tx)$.

1.2. The method is to set $\widehat{K}(u) = Au^4 u_x + Bu^2 u_{3x} + Cuu_x u_{2x} + Du_x^3 + Eu_{5x}$, calculate $(\partial/\partial t)(\widehat{K}(u))$ and $(\partial/\partial s)(K(u))$ by the method indicated after (1.19), then equate coefficients to determine A, B, C, D, E. The answer is
$$\frac{1}{2}u^4 u_x + u^2 u_{3x} + 4uu_x u_{2x} + u_x^3 + \frac{3}{5}u_{5x}.$$

1.3. From $\frac{\partial B}{\partial y} + [B, P] = 0$ we get $3u_{yy} + u_{4x} + 6(uu_x)_x = 0$. This is called the *Boussinesq equation*.

Chapter 2

2.1. Thinking of $\partial \circ f$ as an operator on functions gives $(\partial \circ f)g = \partial(fg) = (f\partial)g + (\partial f)g$. Similarly, $(\partial^n \circ f)(g)$ is the nth order derivative of the product fg, which can be calculated by the Leibniz rule:
$$(\partial^n \circ f)(g) = \partial^n(fg)$$
$$= \sum_{k \geq 0} \binom{n}{k}(\partial^k f)(\partial^{n-k} g).$$

2.2. The answer is
$$\sum_{l=0}^{\infty} h_l \partial^{\alpha+\beta-l}, \quad \text{where} \quad h_l = \sum_{i+j+k=l} \binom{\alpha-i-j}{i} f_j \partial^i g_k,$$
or in other words,
$$\sum_{j,k,k'=0}^{\infty} \binom{\alpha-k}{j}(f_k \partial^j g_{k'}) \partial^{\alpha+\beta-j-k-k'}.$$

2.3. We set $(\partial + x)^{-1} = \sum_{n=1}(-1)^{n-1} a_{n-1} \partial^{-n}$ and multiply out, obtaining $(\partial + x)(a_0 \partial^{-1} - a_1 \partial^{-2} + a_2 \partial^{-3} - \cdots) = 1$. Equating coefficients of ∂^{-n} gives $a_0 = 1$, and $a_n - x a_{n-1} + \partial a_{n-1} = 0$ for $n \geq 1$. Thus $a_0 = 1$, $a_1 = x$, $a_2 = x^2 + 1$, $a_3 = x^3 + 3x$, $a_4 = x^4 + 6x^2 + 3$, etc., and quite generally
$$a_n = \sum_{k=0}^{[n/2]} \frac{n(n-1)(n-2)\cdots(n-2k+1)}{2\cdot 4\cdot 6 \cdots 2k} x^{n-2k}.$$

2.4.
$$\mathrm{ord} L_1 L_2 = \alpha_1 + \alpha_2 \quad \text{and} \quad \mathrm{ord}[L_1, L_2] = \alpha_1 + \alpha_2 - 1.$$

2.5. If l is even then $(P^{l/2})_+ = P^{l/2}$, and therefore $[P, (P^{l/2})_+] = 0$.

2.6.
$$(P^{5/2})_+ = \partial^5 + \frac{5u}{2}\partial^3 + \frac{15}{4} u_x \partial^2 + \left(\frac{25}{8} u_{xx} + \frac{15 u^2}{8}\right)\partial$$
$$+ \frac{15}{16} u_{xxx} + \frac{15}{8} u u_x;$$
$$[P, (P^{5/2})_+] = -\frac{1}{16}(u_{5x} + 10 u u_{xxx} + 30 u^2 u_x + 20 u_x u_{xxx}).$$

2.7.
$$L e^{\xi(\mathbf{x},k)} = (\partial + f_1 \partial^{-1} + f_2 D^{-2} + \cdots)$$
$$\times (1 + w_1 \partial^{-1} + w_x \partial^{-2} + \cdots) e^{\xi(\mathbf{x},k)}$$
$$= (\partial + w_1 + (\partial w_1 + w_2 + f_1) \partial^{-1}$$
$$+ (\partial w_2 + w_3 + f_1 w_1 + f_2) \partial^{-2} + \cdots) e^{\xi(\mathbf{x},k)},$$
so that
$$\partial w_1 + f_1 = 0 \quad \text{and} \quad \partial w_2 + f_1 w_1 + f_2 = 0.$$

No solutions provided to Exercises 2.8–10.

Chapter 3

3.1. One calculates that

$$D_t D_x f \cdot f = 2f f_{xt} - 2f_t f_x \quad \text{and} \quad D_x^4 f \cdot f = 2f f_{4x} - 8 f_x f_{3x} + 6 f_{2x}^2,$$

and thus (3.2) becomes

$$8f f_{xt} - 8 f_t f_x - 2 f f_{4x} + 8 f_x f_{3x} - 6 f_{2x}^2 = 0.$$

We give x weight 1 and t weight 3 and look for homogeneous solutions of degree ≤ 6. Then every polynomial of degree ≤ 3 is a solution. In degree 4, tx is a solution, in degree 5 there is no solution, and in degree 6 there are three solutions

$$-\frac{1}{45} x^6 + \frac{1}{3} t x^3 + t^2, \quad -\frac{1}{3} t x^3 + t^2 \quad \text{and} \quad t^2.$$

3.2. No solution provided.

3.3. Consider the action $B \mapsto [A, B]$, and write $\operatorname{ad}(A)(B) = [A, B]$. Then

$$e^A B e^{-A} = e^{\operatorname{ad}(A)} B,$$

and therefore

$$e^A e^B e^{-A} = \exp(\operatorname{ad}(A) B).$$

But since $\operatorname{ad}(A)(B)$ is a scalar, $\operatorname{ad}(A)^n B = 0$ for $n \geq 2$. It follows that

$$e^A e^B e^{-A} = \exp(B + [A, B]) = e^{[A,B]} e^B.$$

3.4. We have

$$X(p_1, q_1) X(p_2, q_2) = C(p_1, q_1, p_2, q_2) : X(p_1, q_1) X(p_2, q_2) :,$$
$$\text{where} \quad C(p_1, q_1, p_2, q_2) = \frac{(p_1 - p_2)(q_1 - q_2)}{(p_1 - q_2)(q_1 - p_2)}.$$

Here the normal-product notation : : means that we rearrange the order of operators inside the colons, passing all the operations of differentiation to the right and all the multiplications to the left. (This is explained in detail in Chapter 5.) For example,

$$: x_1 \frac{\partial}{\partial x_1} : = : \frac{\partial}{\partial x_1} x_1 : = x_1 \frac{\partial}{\partial x_1}.$$

The expression $C(p_1, q_1, p_2, q_2)$ is symmetric under the transposition $1 \leftrightarrow 2$, so that at first sight it seems that the commutator should vanish. However, on expanding the operator $X(p,q)$ as a sum of monomials in p and q

$$X(p,q) = \sum X_{mn} p^m q^n$$

with coefficients X_{mn}, the commutators $[X_{m_1 n_1}, X_{m_2 n_2}]$ do not have to vanish. This contradiction is resolved in the following way: $\sum_{n=-\infty}^{\infty} x^n$ is a nonzero formal power series in x. But we have

$$\sum_{n=-\infty}^{-1} x^n = \frac{x^{-1}}{1-x^{-1}}, \quad \sum_{n=0}^{\infty} x^n = \frac{1}{1-x},$$

and as rational function

$$\frac{x^{-1}}{1-x^{-1}} + \frac{1}{1-x} = 0.$$

Thus the right way of thinking of $C(p_1, q_1, p_2, q_2)$ is as the formal power series

$$\frac{(1 - \frac{p_2}{p_1})(1 - \frac{q_2}{q_1})}{(1 - \frac{q_2}{p_1})(1 - \frac{p_2}{q_1})}$$

expanded in q_2/p_1 and p_2/q_1, and in this sense it does not vanish. If we write $\delta(x) = \sum_{n=-\infty}^{\infty} x^n$, it has the property that $f(x)\delta(x) = f(1)\delta(x)$ for any formal power series $f(x)$, provided that $f(1)$ is meaningful. Using this we see that

$$[X(p_1, q_1), X(p_2, q_2)] = \frac{\left(1 - \frac{q_1}{p_1}\right)\left(1 - \frac{q_2}{p_2}\right)}{\left(1 - \frac{q_2}{p_1}\right)} \delta\left(\frac{p_2}{q_1}\right) X(p_1, q_2)$$

$$- \frac{\left(1 - \frac{q_1}{p_1}\right)\left(1 - \frac{q_2}{p_2}\right)}{\left(1 - \frac{q_1}{p_2}\right)} \delta\left(\frac{p_1}{q_2}\right) X(p_2, q_1).$$

Thus the vertex operators generate a Lie algebra.

3.5. x_1, $x_2 + \frac{x_1^2}{2}$, $x_2 - \frac{x_1^2}{2}$, $x_3 + x_1 x_2 + \frac{x_1^3}{6}$, $x_3 - \frac{x_1^3}{3}$ and $x_3 - x_1 x_2 + \frac{x_1^3}{6}$. In this case, for $k \in \mathbb{C}$ the only pole is at $k = 0$.

3.6. Setting $x_j = x_j'$ for all j and computing the residue gives $\widetilde{w}_1 = 0$. Next, after we differentiate with respect to x_1', the same calculation gives $\widetilde{w}_2 = 0$, and so on.

3.7. No solution provided.

Chapter 4

4.1. By definition we have $\psi v_1 = \psi|\text{vac}\rangle = 0$ and $\psi v_2 = \psi\psi^*|\text{vac}\rangle = v_1$. Similarly for ψ^*.

4.2. No solution provided.

4.3. Applying Wick's theorem gives $\langle \psi_{m_i}\psi_{m_j}\rangle = 0$ and $\langle \psi^*_{n_i}\psi^*_{n_j}\rangle = 0$, so that the only terms that survive are those of the form

$$\langle \psi_{m_1}\psi^*_{n_{\sigma(1)}}\rangle \cdots \langle \psi_{m_s}\psi^*_{n_{\sigma(s)}}\rangle$$

for a permutation σ. Noting here that the identity permutation $\sigma = \text{id}$ has sign $+1$, we get the answer from the definition of the determinant.

Taking the pairings of the vectors

$$\langle u| = \langle \text{vac}|\psi_{m_1}\cdots\psi_{m_r}\psi^*_{n_1}\cdots\psi^*_{n_s},$$
$$|v\rangle = \psi_{-m'_1}\cdots\psi_{-m'_s}\psi^*_{-n'_1}\cdots\psi^*_{-n'_r}|\text{vac}\rangle$$

for $0 < m_1 < \cdots < m_r$, $0 < n_1 < \cdots < n_s$ and $0 < m'_1 < \cdots < m'_s$, $0 < n'_1 < \cdots < n'_r$, and working as above, neglecting the sign, we get

$$\det\left(\langle\psi_{m_i}\psi^*_{-n'_j}\rangle\right)\det\left(\langle\psi^*_{n_k}\psi_{-m'_l}\rangle\right) = \pm\prod_{i=1}^{r}\delta_{m_i n'_i}\prod_{k=1}^{s}\delta_{n_k m'_k}.$$

It follows form this that the pairing is nondegenerate.

Chapter 5

5.1. Relation (5.10) gives

$$[H_m, H_n] = \sum_j [H_m, \psi_{-j}\psi^*_{j+n}] = \sum_j (\psi_{-j+m}\psi^*_{j+n} - \psi_{-j}\psi^*_{j+m+n}).$$

The false calculation goes as follows: if we remove the brackets in each term of the sum and renumber the first summands by $j \mapsto j+m$, each becomes equal to a second summand, so that they all cancel, to give 0.

Instead, rewrite the terms inside the bracket as

$$:\psi_{-j+m}\psi^*_{j+n} - \psi_{-j}\psi^*_{j+m+n}: + \delta_{m+n,0}(\theta(j<m) - \theta(j<0)).$$

After this, the sum of the normal product parts acts on each vector as a finite sum, so that the renumbering $j \mapsto j+m$ is permissible and the sum gives 0. The remaining constant terms give the contribution $m\delta_{m+n,0}$.

5.2. $x_4 + \frac{1}{2}x_2^2 - \frac{1}{2}x_2x_1^2 - \frac{1}{8}x_1^4$.

5.3. In the expansion $e^{H(\mathbf{x})} = \sum f_j(\mathbf{x})a_j$, the only a_j which have $\langle l|a_j|u\rangle \neq 0$ are those of energy $-d + l^2/2$. Therefore the corresponding coefficients $f_j(\mathbf{x})$ are all of weight $d - l^2/2$.

5.4. For each n separately, a term ψ_n (respectively ψ_n^*) appearing in (4.11) contributes zq^{-n} (respectively $z^{-1}q^{-n}$) to the character.

5.5. The character of $\mathbb{C}[x_1, x_2, x_3, \ldots]$ is given by
$$\sum_{m_1,m_2,\ldots \geq 0} q^{m_1+2m_2+\cdots} = \prod_{n=1}^{\infty} \sum_{m_n=0}^{\infty} q^{nm_n} = \prod_{n=1}^{\infty} (1-q^n)^{-1}.$$

We have that $\langle l|$ has charge l and energy $l^2/2$. Now because we can identify $\mathcal{F}_l$ with $z^l \mathbb{C}[x_1, x_2, x_3, \ldots]$, the character of $\mathcal{F}$ can be computed as in the question. For the second part, it is enough to put together the formulas of Exercises 5.3 and 5.4 and replace z by $-zq^{-1/2}$.

Chapter 6

6.1. We omit the first half. For the second half, we see that the section $z = 0$ of (6.2) is contained in the orbit, for example by using elements
$$g_1 = \begin{pmatrix} a & b & \\ b & a & \\ & & 1 \end{pmatrix}.$$

Then G contains the rotation around the x axis given by
$$g_2 = \begin{pmatrix} 1 & & \\ & \cos\theta & -\sin\theta \\ & \sin\theta & \cos\theta \end{pmatrix},$$

which gives the locus (6.2).

No solution provided to Exercises 6.2–3.

Chapter 7

7.1. It is enough to prove that for $X, Y \in \mathfrak{sl}_2$, the elements $A(t) = Xt^m$ and $B(t) = Yt^n \in \widehat{\mathfrak{sl}_2}$ satisfy $\omega(A, B) = m\delta_{m+n,0}\text{Tr}(XY)$.

Chapter 8

8.1. For example, suppose that the sphere is tangent to a plane at the south pole, and consider the stereographic projection of the sphere away from the north pole onto this plane.

8.2. No solution provided.

8.3. Let W be an m dimensional vector subspace of V and $M_W = (v_{ij})$ (for $1 \leq i \leq m$ and $1 \leq j \leq N$) a frame of W. The necessary and sufficient condition for a vector $(x_1, \ldots, x_N) \in V$ to belong to W is that the $(m+1) \times N$ matrix obtained by adding this vector to the rows of M has rank m. Thus for any $(\beta_1, \ldots, \beta_{m+1})$ we must have

$$\sum_{j=0}^{m+1} (-1)^j x_{\beta_j} v_{\beta_1, \ldots, \beta_{j-1}, \beta_{j+1}, \ldots, \beta_{m+1}} = 0.$$

This set of $\binom{N}{m+1}$ hyperplanes of V is uniquely determined by the Plücker coordinates of W, and W is contained in their intersection. If we can prove that the above linear system of equations in the x_i has solution space of dimension $\leq m$ then W equals this intersection, and is uniquely determined by its Plücker coordinates.

For this, suppose that $v_{\alpha_1, \ldots, \alpha_m}$ is one of the nonzero Plücker coordinates. Then set $(\beta_1, \ldots, \beta_{m+1}) = (\alpha_1, \ldots, \alpha_m, i)$ for $i \neq \beta_j$, and write out the above equation. We get

$$\sum_{j=0}^{m} (-1)^j x_{\alpha_j} v_{\alpha_1, \ldots, \alpha_{j-1}, \alpha_{j+1}, \ldots, \alpha_m, i} + (-1)^{m+1} x_i v_{\alpha_1, \ldots, \alpha_m} = 0.$$

Renumbering the Plücker coordinates gives

$$x_i v_{\alpha_1, \ldots, \alpha_m} - \sum_{j=1}^{m} x_{\alpha_j} v_{\alpha_1, \ldots, \alpha_{j-1}, i, \alpha_{j+1}, \ldots, \alpha_m} = 0.$$

Partitioning the indices as $\{1, \ldots, N\} = \{\alpha_1, \ldots, \alpha_m\} \cup \{\alpha_{m+1}, \ldots, \alpha_N\}$ and considering the above equation for each $i = \alpha_{m+1}, \ldots, \alpha_N$, we see that the $(N-m) \times N$ matrix of coefficients of the above system of equations has rank $N - m$. Indeed, the minor formed by its columns numbered $\alpha_{m+1}, \ldots, \alpha_N$ is $(v_{\alpha_1, \ldots, \alpha_m})^{N-m} \neq 0$.

8.4. Let $\sum c_{\alpha_1, \ldots, \alpha_m} v_{\alpha_1, \ldots, \alpha_m} = 0$ be a linear dependence relation; we need to prove that all the coefficients $c_{\alpha_1, \ldots, \alpha_m} = 0$. Consider the m dimensional subspace spanned by the first m standard vectors

$\mathbf{v}_i = (v_{ij})$ for $i = 1, \ldots, m$ with $v_{ij} = \delta_{ij}$. Then the Plücker coordinates of this subspace are $v_{1,\ldots,m} = 1$ and $v_{\alpha_1,\ldots,\alpha_m} = 0$ for $\{\alpha_1, \ldots, \alpha_m\} \neq \{1, \ldots, m\}$; therefore $c_{1,\ldots,m} = 0$.

8.5. Perform the Laplace expansion of the determinant

$$\begin{vmatrix} v_{1\beta_1} & \cdots & v_{1\beta_{m+1}} & 0 & \cdots & 0 \\ \cdot & \cdots & \cdot & \cdot & \cdots & \cdot \\ v_{m\beta_1} & \cdots & v_{m\beta_{m+1}} & 0 & \cdots & 0 \\ v_{1\beta_1} & \cdots & v_{1\beta_{m+1}} & v_{1\alpha_1} & \cdots & v_{1\alpha_{m-1}} \\ \cdot & \cdots & \cdot & \cdot & \cdots & \cdot \\ v_{m\beta_1} & \cdots & v_{m\beta_{m+1}} & v_{m\alpha_1} & \cdots & v_{m\alpha_{m-1}} \end{vmatrix},$$

along the top m rows.

Chapter 9

9.1. We write $\bigwedge V_N$ for the exterior algebra generated by V_N. As a vector space, it has dimension 2^{2N}. We can identify it with the subalgebra of $\mathcal{A}_N$ generated by the ψ_i. The ψ_i and ψ_i^* correspond respectively to the following linear operators on $\bigwedge V_N$:

$$\psi_i: \psi_{i_1} \wedge \psi_{i_2} \wedge \psi_{i_3} \wedge \cdots \mapsto \psi_i \wedge \psi_{i_1} \wedge \psi_{i_2} \wedge \psi_{i_3} \wedge \cdots;$$
$$\psi_i^*: \psi_{i_1} \wedge \psi_{i_2} \wedge \psi_{i_3} \wedge \cdots \mapsto [\psi_i, \psi_{i_1}]_+ \psi_{i_2} \wedge \psi_{i_3} \wedge \cdots$$
$$- [\psi_i, \psi_{i_2}]_+ \psi_{i_1} \wedge \psi_{i_3} \wedge \cdots + \cdots.$$

This correspondence gives an isomorphism of $\mathcal{A}_N$ with the algebra of all $2^{2N} \times 2^{2N}$ matrixes.

The fact that the centre of the algebra contains only scalar multiples of the identity can also be proved as follows. Let a be a central element of $\mathcal{A}_N$. Because a commutes with $H_0 = \sum_{i>0} \psi_{-i}\psi_i^* - \sum_{i<0} \psi_i^*\psi_{-i}$, it follows that it has charge 0. Write

$$a = \sum c(m_1, \ldots, m_r, n_1, \ldots, n_r) \psi_{m_1} \cdots \psi_{m_r} \psi_{n_1}^* \cdots \psi_{n_r}^*$$

(the term with $r = 0$ is constant); here we have assumed that the set of terms $\psi_{m_1} \cdots \psi_{m_r} \psi_{n_1}^* \cdots \psi_{n_r}^*$ on the right-hand side is linearly independent. Now consider $[\psi_{m_1} \cdots \psi_{m_r} \psi_{n_1}^* \cdots \psi_{n_r}^*, \psi_i]$; this is nonzero only if $i = -n_l$ for some l, and then it equals

$$(-1)^{r-l} \psi_{m_1} \cdots \psi_{m_r} \psi_{n_1}^* \cdots \psi_{n_{l-1}}^* \psi_{n_{l+1}}^* \cdots \psi_{n_r}^*.$$

These terms are linearly independent, so that each of their coefficients must be 0. Now the assertion follows from considering also the commutator with ψ_i^*.

9.2. Suppose that $\dim V(|u\rangle) = N + r$. For suitable $g \in \mathbf{G}$, we can write
$$T_g V(|u\rangle) = \bigoplus_{i>-r} \mathbb{C}\psi_i$$
$$= V(\psi_{-r+1/2} \cdots \psi_{-1/2}|\text{vac}\rangle).$$

By (9.5) and the remark following it, we obtain
$$V(|u\rangle) = V(g\psi_{-r+1/2} \cdots \psi_{-1/2}|\text{vac}\rangle),$$
but since g preserves charge, this contradicts the assumption that $|u\rangle$ has charge 0.

9.3. No solution provided.

9.4. (a) $-x_4 - x_2^2/2 + x_2 x_1^2/2 + x_1^4/8$;
(b) $x_5 x_1 - x_3^2 - x_3 x_1^3/3 + x_1^5/45$;
(c) $x_6 x_1 - x_4 x_3 + x_4 x_2 x_1 - x_4 x_1^3/6 - x_3^2 x_1/2 - x_3 x_2^2/2 - x_3 x_2 x_1^2/2 - x_3 x_1^4/24 + x_2^3 x_1/6 - x_2^2 x_1^3/12 + x_2 x_1^5/24 + x_1^7/144$.

Chapter 10

10.1. Use the fact that $\psi_{1/2}^* \psi_m(\mathbf{x}) = p_{-m-1/2}(\mathbf{x}) - \psi_m(\mathbf{x})\psi_{1/2}^*$ and
$$h_{-m-1/2,-n-1/2}(\mathbf{x}) = -h_{-m+1/2,-n-3/2}(\mathbf{x})$$
$$+ p_{-m-1/2}(\mathbf{x}) q_{-n+1/2}(\mathbf{x}).$$

10.2. No solution provided.

10.3.
$$(\chi_{(3)}, \chi_{(2,1)}, \chi_{(1^3)}) = (x_3, x_2 x_1, x_1^3/6) A,$$
$$\text{where } A = \begin{pmatrix} 1 & -1 & 1 \\ 1 & 0 & -1 \\ 1 & 2 & -1 \end{pmatrix}.$$

Transforming $(\partial_3, \partial_2 \partial_1, \partial_1^3)$ under ${}^t A^{-1}$ gives
$$(\chi_{(3)}(\partial_\mathbf{x}), \chi_{(2,1)}(\partial_\mathbf{x}), \chi_{(1^3)}(\partial_\mathbf{x})).$$

10.4. The sum (or alternatively, the difference) of the Plücker relation determined by the set consisting of the empty Young diagram of charge -1 and the Young diagram $(1,1,1,1)$ of charge 1 and the Plücker relation determined by the set consisting of the empty Young diagram of charge -1 and the Young diagram $(2,1,1)$ of charge 1.

Bibliography

This book has concentrated mainly on the results on soliton equations and their symmetries of the Kyoto school surrounding Mikio SATO, work that reached its peak in the years around 1981. The following references are available for the reader interested in taking this further.

M. Sato, Lecture notes taken by N. Umeda (in Japanese), *Kyoto Univ. RIMS Lecture Notes* **5**, 1989

M. Sato, Soliton equations and universal Grassmann varieties, Lecture notes taken by M. Noumi (in Japanese), *Saint Sophia Univ. Lecture Notes* **18**, 1984

E. Date, M. Kashiwara, M. Jimbo and T. Miwa, Transformation groups for soliton equations, *Proc. Japan Acad.* **57A** (1981), 342–7, 387–92; *J. Phys. Soc. Japan* **50** (1981), 3806–12, 3813–18; *Physica* **4D** (1982), 343–65; *Publ. RIMS, Kyoto Univ.* **18** (1982), 1111–19, 1077–110.

E. Date, M. Kashiwara, M. Jimbo and T. Miwa, Transformation groups for soliton equations, in *Nonlinear integrable systems – classical theory and quantum theory (Kyoto, 1981)*, World Scientific, Singapore, 1983, pp. 39–119

M. Jimbo and T. Miwa, Solitons and infinite-dimensional Lie algebras, *Publ. RIMS, Kyoto Univ.* **19** (1983), 943–1001

There are several books treating the background to the theory of solitons, the inverse scattering method, quasiperiodic solutions and so on, and their physics and engineering applications, and we have not gone into these topics here. We mention some books for those who can read Japanese:

T. Tanaka and E. Date, *The KdV equations*, Kinokuniya, Tokyo 1979

R. Hirota, *Solitons by the direct method*, Iwanami, Tokyo 1992

One of the most significant developments in mathematics from the 1970s through to the present (1992) is the progress stimulated by interaction between various areas of mathematics and physics (among these, representation theory, differential equations, algebraic geometry, number theory, operator algebras, topology, combinatorics and so on). The development of soliton theory was one of the first events in this development. Undoubtedly an attractive topic for an essay would be to take soliton theory as a key to understanding the history of mathematics in the final quarter of the 20th century or more; however, this is not something we cannot embark on at present (and maybe it is impossible by definition!). It is not even clear whether this process is going to converge in the final years of the 20th century. Be that as it may, it seems appropriate for us to present the reader kind enough to glance through our book with a rough overview of the present state of the literature.

A first point of view, and a central concern of this book, is that of infinite dimensional Lie algebras. The most important infinite dimensional Lie algebras are the Kac–Moody Lie algebras, and in particular the affine Lie algebras; a basic reference for these is

V. G. Kac, *Infinite dimensional Lie algebras*, third edition, CUP, New York, 1990

The Hamiltonian reduction of Drinfel'd and Sokolov is an alternative point of view, relating soliton theory and affine Lie algebras:

V. G. Drinfel'd and V. V. Sokolov, Lie algebras and equations of Korteweg–de Vries type (in Russian), *Current problems in mathematics*, **24**, 81–180, Itogi Nauki i Tekhniki, Akad. Nauk SSSR, Vsesoyuz. Inst. Nauchn. i Tekhn. Inform., Moscow, 1984, English translation: *J. Sov. Math.* **30** (1985), 1975–2036

The fact that vertex operators provide an action of affine Lie algebras on soliton solutions has played a central role in this book; the following papers treat vertex operators from the point of view of representation theory:

J. Lepowsky and R. L. Wilson, Construction of the affine Lie algebra $A_1^{(1)}$, *Comm. Math. Phys.* **62** (1978), 43–53

I. B. Frenkel and V. G. Kac, Basic representations of affine Lie algebras and dual resonance models, *Invent. Math.* **62** (1980/81), 23–66

These papers only treat level 1 representations. Treating general levels requires completely different ideas:

M. Wakimoto, Fock representations of the affine Lie algebra $A_1^{(1)}$ *Comm. Math. Phys.* **104** (1986), 605–9

B. L. Feigin and E. V. Frenkel, Representations of affine Kac–Moody algebras and bosonization, in *Physics and mathematics of strings, V. Knizhnik memorial volume*, L. Bink and others (eds.), World Scientific, Singapore, 1990, pp. 271–316

Conformal field theory takes the representation theory of infinite dimensional Lie algebras as its backdrop, and constructs a conformally invariant quantum field theory consisting of massless particles:

A. A. Belavin, A. M. Polyakov and A. B. Zamolodchikov, Infinite conformal symmetry in two-dimensional quantum field theory, *Nuclear Phys.* B **241** (1984), 333–80.

Here the vertex operators relating the different irreducible representations of infinite dimensional Lie algebras play the role of operators on fields.

A. Tsuchiya and Y. Kanie, Vertex operators in conformal field theory on $\mathbb{P}^1$ and monodromy representations of braid group, in *Conformal field theory and solvable lattice models (Kyoto, 1986)*, Adv. Stud. Pure Math., **16**, Academic Press, Boston, MA, 1988, pp. 297–372

One can obtain a theory of massive fields as a deformation of conformal field theory, and this is nothing other than a quantisation of soliton theory.

A. B. Zamolodchikov, Integrable field theory from conformal field theory. in *Integrable systems in quantum field theory and statistical mechanics*, Adv. Stud. Pure Math., **19**, Academic Press, Boston, MA, 1989, pp. 641–74

T. Eguchi and S.-K. Yang, Deformations of conformal field theories and soliton equations, Phys. Lett. B **224** (1989), 373–8

B. Feigin, and E. Frenkel, Free field resolutions in affine Toda field theories Phys. Lett. B **276** (1992), 79–86

A second point of view is that of algebraic geometry. Among the solutions of soliton equations, this book has treated solutions as rational or exponential functions. The following paper gives a definitive treatment in general form of the old idea that solutions in terms of elliptic functions can be constructed from theta functions of an algebraic curve.

I. M. Krichever, Methods of algebraic geometry in the theory of nonlinear equations (in Russian), *Uspehi Mat. Nauk* **32**:6 (198) (1977), 183–208, English translation: *Russian Math. Surveys* **32** (1977), 185–213

For this, the Japanese reader may see the book by Date and Tanaka cited above. The inverse problem is to know when a theta function that provides the tau function for the KP equation corresponds to an algebraic curve; Novikov's conjecture on this subject was solved affirmatively by T. Shiota:

T. Shiota, Characterization of Jacobian varieties in terms of soliton equations, *Invent. Math.* **83** (1986), 333–82

The following paper obtains a generalisation of the theory of theta functions in terms of the conformal field theory problem of constructing quantum field theory over a Riemann surface:

A. Tsuchiya, K. Ueno and Y. Yamada, Conformal field theory on universal family of stable curves with gauge symmetries, in *Integrable systems in quantum field theory and statistical mechanics*, Adv. Stud. Pure Math., **19**, Academic Press, Boston, MA, 1989, pp. 459–566

A third point of view is that of an integrable system with infinite degrees of freedom. We have already mentioned conformal field theory. Important work previous to this is the theory of correlation functions of the Ising model, Painlevé equation and isomonodromic deformation.

T. T. Wu, B. M. McCoy, C. A. Tracy and E. Barouch, Spin–spin correlation functions for the two-dimensional Ising model: exact theory in the scaling region, *Phys. Rev.* B **13** (1976), 316–74

M. Sato, T. Miwa and M. Jimbo, Holonomic quantum fields. Publ. RIMS, Kyoto Univ.: I, **14** (1978), 223–67; II, The Riemann–Hilbert problem. **15** (1979), 201–78; III, **15** (1979), 577–629. IV, **15** (1979), 871–972; V, **16** (1980), 531–84

The KdV equation turns up also in quantum gravity related to the moduli space of algebraic curves.

M. Kontsevich, Intersection theory on the moduli space of curves and the matrix Airy function, *Comm. Math. Phys.* **147** (1992), 1–23

The Painlevé equations appear also in topological quantum field theory, relating to conformal field theory and soliton theory:

S. Cecotti, P. Fendley, K. Intriligator, and C. Vafa, A new supersymmetric index, *Nuclear Phys.* B **386** (1992), 405–52

A fourth viewpoint is that of quantum groups. Quantum groups were introduced independently by Drinfel'd and Jimbo. We give the following references:

V. G. Drinfel'd, Quantum groups, in *Proceedings of the ICM (Berkeley, 1986)*, Amer. Math. Soc., Providence, RI, 1987, pp. 798–820

M. Jimbo, Quantum groups and the Yang–Baxter equation (in Japanese), Springer, Tokyo, 1990

Quantum groups arose out of solvable lattice models and the quantum inverse scattering method:

R. J. Baxter, Exactly solved models in statistical mechanics, Academic Press, London–New York, 1982, reprinted 1989

E. K. Sklyanin, L. A. Takhtadzhyan and L. D. Faddeev, Quantum inverse problem method. I (in Russian) *Teoret. Mat. Fiz.* **40** (1979), 194–220

In addition to solvable lattice models, quantum groups turn up in many areas, such as invariants of low dimensional topology, braid representations of conformal field theory, symmetries of solvable massive field theory, and so on. Finally, we give some references to a more recent topic, the q-deformation of vertex operators.

I. B. Frenkel and N. Yu. Reshetikhin, Quantum affine algebras and holonomic difference equations, *Comm. Math. Phys.* **146** (1992), 1–60

F. A. Smirnov, Dynamical symmetries of massive integrable models, in *Infinite analysis (Kyoto, 1991)*, Adv. Ser. Math. Phys., **16**, World Scientific, River Edge, NJ, 1992, pp. 813–37 and 839–58

B. Davies, O. Foda, M. Jimbo, T. Miwa and A. Nakayashiki, Diagonalization of the XXZ Hamiltonian by vertex operators, *Comm. Math. Phys.* **151** (1993), 89–153

Index

affine Lie algebra $\widehat{sl_2}$, 61, 64, 65, 104
annihilation operator, 33–41, 45, 50, 59, 80
anticommutator, 34, 54, 77
  bracket, 46

bilinear identity, 28–29, 58–60, 76, 80, 81, 88
Boson, 33–35, 43, 46, 63
  –Fermion correspondence, 43, 50, 57–58, 81, 82, 84
Bosonic
  algebra, 32
  commutation relations, 46
  Fock space, 33, 47, 49, 52, 57, 88
  normal product, 45

canonical
  anticommutation relation, 34, 36, 76
  commutation relation, 32–34, 46
Cauchy's identity, 44, 51
character, 52, 84, 99
  polynomial, 76, 82–85, 87–93
charge, 37, 39, 40, 47, 50, 55, 57, 77, 79–82, 84, 86–89, 99, 101, 102
charged Young diagram, 86
Clifford
  algebra, 34–35, 41, 45, 76, 87
  conjugation, 78
  group, 76–78
cocycle condition, 56, 60
colon notation, 45
commutator, 41, 54, 56, 96
  bracket, 4, 8, 31, 46, 54–56, 64
complex projective line, 67
complex projective space, 67
creation operator, 33, 34, 36–38, 40–41, 50

differential

  equation, 3, 5, 32, 93
  polynomial, 6–7
Dirac's sea, 37
dual wave function, 58

eigenvalue, 8, 9
electric charge, *see* charge
energy, 37, 39, 40, 47, 52, 77, 99
exponent, 4, 21, 23

Fermion, 32, 34–36, 43
Fermionic
  Fock space, 36, 49
Fock
  representation, 35, 37, 41
  space, 37, 46–47, 76, 80
frame, 68, 100

generating function, 43, 46, 50–52, 57
Grassmannian, 66, 68, 69, 75, 76, 80
group, 1, 2, 5
  action, ix, 3, 53

Heisenberg algebra, 33–35, 64
higher order KdV equation, 9, 13, 23
Hirota
  bilinear method, viii, 19, 27, 92–93
  derivative, ix, 19–24
  equation, 19–23, 53, 60, 61, 65, 88, 90, 92–93
homogeneous, 91, 96
  coordinates, 69–70
  differential polynomial, 7
  polynomial, 44

infinite dimensional
  Grassmannian, ix, 76, 88
  group, ix, 53
  Lie algebra, 53, 65, 104
  symmetries, ix, 5, 14, 32

Index

infinitesimal
  generator, 3–5, 7
  transformation, 1, 3–5, 9, 15, 25, 32, 59–61, 63–65
integrable system, viii–ix, 22, 105

KdV equation, vii–ix, 1, 5–7, 9, 11, 15, 19–24, 31, 32, 61, 63–65, 105
KdV hierarchy, 15, 23–25, 27, 61
KP equation, ix, 27, 32, 53, 57, 59–60, 105
KP hierarchy, 16, 17, 27–29, 31, 58, 60, 61, 63, 76, 90, 92

Lax form, 8–10, 13, 59
level, 65, 104
Lie algebra, 3–5, 53, 54, 56–58, 60, 63–65, 80, 97
linear differential equation, vii, viii, 8, 61

Maya diagram, 35–37, 82, 85–87, 89–90

nonlinear differential equation, viii, ix, 17
normal product, 44–46, 55, 96, 98

operator algebras, 32–33, 44, 57, 103
orbit, 53–54, 58, 60, 64, 65, 76, 79, 80, 90, 99

parabolic subgroup, 68
Plücker, 69
  coordinates, 69–72, 75, 76, 92, 100–101
  embedding, 70
  relations, 66, 71–76, 80–82, 88, 90–93, 102
Plücker relations, 102
point at infinity, 67
projective
  geometry, 66
  line, 67
  space, 67, 68, 70, 72
pseudodifferential operator, 11–13, 16–18, 61

quantum
  field theory, 43, 104–105
  group, 106

representation, 33, 35, 57, 58, 65, 82, 84, 103, 104

signed Maya diagram, 89–90
soliton, viii, 32, 103
  equation, viii, 19, 65, 103, 105
  solution, 21–22, 25–27, 31, 60, 61, 63, 104
spectral variable, 8
symmetry, ix, 1–2, 4–7, 9, 14–15, 23–24, 32, 53

tau function, 17, 22, 25, 27–31, 53, 58–60, 63, 65, 76, 88, 90–93, 105
Toda lattice, viii
transformation group, ix, 1–2, 5, 53, 58, 61, 63

vacuum
  expectation value, 40–41, 43
  orbit, 80
  state, 33, 34, 36–39, 41, 46, 58, 64, 65, 76–77, 79–81, 90
  vector, 41, 53, 80
vertex operator, 19, 24, 25, 27, 57, 60, 63–64, 97, 104, 106

wave function, 58, 60
Wick's theorem, 41, 44, 59, 85, 88, 98

Young diagram, 82–88, 91, 102